PLANT LIFE ON
EAST ANGLIAN HEATHS

PLANT LIFE ON
EAST ANGLIAN HEATHS

BEING OBSERVATIONAL AND EXPERIMENTAL STUDIES OF THE VEGETATION OF BRECKLAND

By

E. PICKWORTH FARROW, M.A., D.Sc.

CAMBRIDGE
AT THE UNIVERSITY PRESS
1925

CAMBRIDGE UNIVERSITY PRESS
Cambridge, New York, Melbourne, Madrid, Cape Town,
Singapore, São Paulo, Delhi, Tokyo, Mexico City

Cambridge University Press
The Edinburgh Building, Cambridge CB2 8RU, UK

Published in the United States of America by Cambridge University Press, New York

www.cambridge.org
Information on this title: www.cambridge.org/9781107605107

First published 1925
First paperback edition 2011

A catalogue record for this publication is available from the British Library

ISBN 978-1-107-60510-7 Paperback

PREFACE

THIS book is an account of some observations, problems and experimental work relating to the Ecology of the vegetation of the East Anglian heath district known as the "Breck Country."

The Breck Country lies in the low rainfall area of East Anglia—the annual average being in the neighbourhood of 22·5 inches (about 560 mm.)—and the upper layers of soil are sandy and very porous. The district represents the nearest approach to Steppe conditions in Great Britain, and thus it was thought that work on the vegetation of this district might lead to interesting results. It should be noted that the dominance of the common heather or ling (*Calluna vulgaris*), a plant of oceanic and suboceanic climates, separates the vegetation very decidedly from that of true continental steppe.

It is difficult to describe the complex natural phenomena of vegetation in words and in this book photographs are largely employed. They may give slight approach to actual acquaintance-knowledge of the district and reduce the necessity for a large amount of verbal description.

It is hoped that the relatively large number of photographs will be useful and interesting to the reader.

The experimental method has been employed whenever possible. It is felt that ecological work is too often content with mere description, and that actual experiment *in the field* should be far more extensively used in the endeavour to obtain real knowledge of the determining causes of plant groupings. Until ecology becomes an experimental science its conclusions must remain largely uncertain and relatively superficial.

The very great importance of the biotic factors of the environment in bringing about the disappearance of various former vegetation units and thereby permitting other forms of vegetation to replace them, emphasised in the following pages, has been found to be of widespread application to vegetation in general and it has been a pleasure to the writer to have this work referred to in relation to the vegetation of areas so widely separated as Central and South-Western Australia, Central and North America, South Africa, Japan, the mountains of North Wales, and the Swiss Alps.

Lantern slides of any of the photographs and diagrams in this book may be obtained from Messrs Flatters & Garnett, 309 Oxford Road, Manchester, price 1s. 6d. each.

Most of the book has already appeared in instalments in *The Journal of Ecology*, 1915–1924, and the thanks of the writer are due to the Editor of that Journal and to the British Ecological Society for permission to reprint it.

The writer wishes to express his sincere thanks to Mr (now Professor) R. S. Adamson who suggested that work on the Breckland Heaths was needed, to Dr (now Professor) C. E. Moss for encouragement in the initial stages and for suggestions relating to the valley fen woodlands, to Prof. J. E. Marr, F.R.S., for geological references and for arousing interest in various related topics and to Mr (now Dr) W. Watson for naming lichen specimens.

The writer is deeply indebted to Mr A. G. Tansley, F.R.S., for much constant encouragement and frequent advice during the progress of the work and for very kindly reading the manuscript and suggesting many improvements therein, and also to Prof. F. W. Oliver, F.R.S., who aroused his interest in plant ecology and from whom at Blakeney Point he has learnt much, especially with regard to general outlook on the phenomena associated with vegetation.

E. P. F.

SPALDING,

September, 1924.

CONTENTS

CHAPTER I

General Description of Breckland and Its Vegetation

CHAPTER II

Factors Relating to the Relative Distributions of Calluna *Heath and* Grass Heath *in Breckland*

CHAPTER III

General Effects of Rabbits on the Vegetation

Contents

LIST OF ILLUSTRATIONS.

CHAPTER I.

GENERAL DESCRIPTION OF BRECKLAND AND ITS VEGETATION.

SYNOPSIS.

I.   INTRODUCTION.

The Breck Country[1] lies in the north-west corner of Suffolk and the south-west corner of Norfolk (see Figs. 1 and 2), beyond the eastern border of the Fenland, and has an area of about 400 square miles and an altitude varying from 20 to 150 feet above sea level.   The underlying rock is chalk, and most of the heaths are on sand-covered chalk spurs of the East Anglian Heights.

II.   GENERAL DESCRIPTION OF BRECKLAND.

The ground in Breckland usually rises gradually from the river valleys, making the apparent altitudes greater than the actual altitudes and views of extensive areas are sometimes obtained[2].   Much of the district consists of sandy plains of moderate elevation.   The district shows fine transitions from dry sandy heath on the upper portions to woodland and fen in the river valleys but no natural woodland has been found on the higher parts. Most of the chalk is covered by boulder clay, which however is never very thick.   The geology of the district is much obscured by layers of gravel and sand which are spread over both chalk and boulder clay alike, and owing to these coverings of drift and sand the characteristic features of a chalk tract are not much in evidence—there are no deep valleys and no sharp escarpments.   The presence of the boulder clay is revealed by its existence in pits which have been dug through the layers of sand and gravel; formerly some of the sandy brecks were cultivated and clay used to be removed from these pits for mixing with the sandy surface soil to improve it.

---

[1] So called because in the past, after the passing of enclosure acts, portions of ancient common were ploughed or broken up and were termed "brecks" (breck = broken).

[2] **Clarke, W. G.** "Some Breckland Characteristics." *Trans. Norfolk and Norwich Naturalists Society,* **8,** 1908.

The annual rainfall is 22·5 inches, which is one of the lowest in Britain, approaching the low average of the Fenland. Some portions of the district are so barren as to present a type of scenery not met with elsewhere in England, the dry sandy areas being of a type which is unique in this country. Heathland meres form a remarkable feature of the district; these have no visible inlet or outlet and occur only in those localities where the chalk comes near the surface or is but thinly covered by sandy drift.

Fig. 1. Map of East Anglia, showing the position of Breckland and its relation to the ancient bay of the Wash.

The original bay of the Wash, hollowed out in Jurassic clays, once extended almost to Mildenhall (Figs. 1 and 2), and to-day a strong current of wind-blown sand may often be seen at Hunstanton going south in the direction of Lynn along the present edge of the Wash. Possibly some of the sand covering Breckland may be derived from ancient postglacial sand dunes on the former south-east border of the Wash before the alluvium of the present Fenland was deposited in the ancient bay of the Wash (Fig. 1). The prevailing direction of the winds here is south-west, and this would disperse the sand from Mildenhall in the direction of Thetford, which is in

Fig. 2. Map of Breckland and the surrounding district.

fact its actual distribution. In historic times great storms of wind-blown sand have occurred, and smaller sandstorms used to be frequent in the district before the planting of long pine belts which break the force of the wind. The view that the Breckland sands may be derived from ancient sand dunes on the south-east edge of the original bay of the Wash is also supported by the occurrence in Breckland, and especially in the Mildenhall district, of coastal dune plants like *Carex arenaria* and *Phleum arenarium* in a habitat which has long been inland and abnormal for them. Various littoral insects, birds and shells also occur here, and the presence of these plants and animals characteristic of coastal sand dunes in what has long been an inland area, now far from the sea, is one of the most interesting features of Breckland.

The district shows numerous evidences of man's prehistoric activities, in the form of flint implements of various Palaeolithic and Neolithic types, tumuli, burial grounds and linear defensive earthworks. The long occupation of the district by man and his domesticated herds has probably reacted considerably on its natural vegetation and may have brought about the degeneration of possible primitive woodland to practically pure heath associations. But though in early times man apparently inhabited this area more extensively than the surrounding country, yet to-day the reverse holds good and the more sterile portions of the district are very little affected by man's activities, this helping to make them relatively more interesting from an ecological point of view. Many of the most sterile sandy areas have probably never been cultivated at all, and the commons are probably survivals of the vast tracts of land formerly held by the community and gradually lost by successive enclosures.

Most of the sandy upper parts are now occupied by heather, bracken and dwarf grass-heath vegetation. Many pine plantations and belts have been planted on these parts in relatively recent times, but no remnant of natural woodland occurs here though luxuriant tree growth is found in the valleys. The transition from dry sandy heath to woodland often takes place within very short distances; in the case of the valley shown in Photo. 1, there were very few trees fifty years ago, but now there are many, the trees being apparently able to colonise the edge of the *Calluna* heath. Another of the various types of valley zonation, some of which are probably in a state of flux, is shown in Photo. 2. Sometimes seedling trees occur in very slight hollows in the heather heath, probably due to the water table being nearer the surface in these places; apparently very slight differences in water table may be associated with considerable differences in the vegetation. A typical portion of one of the characteristic grass-heath areas is shown in Photos 3 and 4; this association will be described later. These grass-heath associations alternate with *Calluna* heaths in the district. There are many rabbit burrows in these grass-heaths.

Some cultivation is still carried on near the villages, but agriculturally

PLATE I

Photo. 1. Zonation on the edge of Cavenham Heath : dry sandy heath covered by heather (*Calluna vulgaris*) on upper portion towards the right ; on the left, in the valley, *Betula* and *Pinus* seedlings with adult trees ; and *Salix repens*. (See p. 4.)

Photo. 2. Zonation on Cavenham Heath. *Pteris aquilina* on drier well-drained upper portion towards the left ; *Erica tetralix* (rounded cushions) and *Carex arenaria* in centre ; *Salix repens* towards right ; pine plantation in background. (See p. 4.)

the soil is very poor and much of the land once farmed has passed out of cultivation. It will not grow wheat profitably, and though rye and barley can be grown the crops are very light; buckwheat is grown frequently but is used chiefly as food for game and is not harvested; potatoes of good quality and very free from disease are raised, but here again the crop is light; lucerne grows well when once thoroughly established, and lupins often grow fairly well and are sometimes ploughed into the sandy soil as green manure, but on the whole the land is too dry for successful agriculture.

Considerable quantities of manure have been applied to some of the fields in the valleys, but when attempts are made to cultivate the sandy upper areas comparatively small quantities of manure are applied to them.

These facts are interesting on the general theory of limiting factors for on the drier upper areas water-supply is anyhow a severe (i.e. heavy and long continued) limiting factor to the growth of vegetation and under these conditions it is not worth while to add manure the supply of which is not a limiting factor, while in the valleys the available amounts of various manurial constituents (and not water supply) are limiting factors to the growth of vegetation and under these other conditions it is worth while to manure the fields in the valleys comparatively heavily.

Cavenham Heath is fairly typical of the Breckland heaths in its topography and vegetation, and most of the detailed work up to the present has been done on it.

III. General Description of Cavenham Heath.

Cavenham Heath, the position of which is indicated in the accompanying maps (Figs. 1 and 2), is bounded on the north by the river Lark, on the east and partly on the south by the road from Icklingham to Cavenham, and on the west by pine plantations and a small stream running from these down to the river Lark. It consists on the whole of a plain partly intersected by two slight valleys. The northern portions slope gradually down to the valley of the Lark, and the extreme southern portions down towards Cavenham. The general altitude is about 80 feet. The underlying rock is chalk, covered by Quaternary gravel and sand deposits, though towards the river valleys the surface layers of the soil become alluvial in character. The soil is very dry and porous, and the vegetation consists chiefly of *Calluna* with areas of grass heath. The *Calluna* heath tends to occupy the more central portions of the upper areas and to be surrounded by stretches of grass-heath, but this arrangement is often modified by the existence of large tracts dominated by bracken, while *Carex arenaria* also occupies extensive areas which are shown in the map (Fig. 3). There are three tree plantations on the heath, two consisting of *Pinus sylvestris* and the third chiefly of *Quercus sessiliflora*. In the damp alluvial valleys of the Lark and of its tributary which runs east of the Icklingham-Cavenham road, valley-fen woods occur (see Photos 10–12).

In Fig. 3, showing the distribution of the various types of vegetation,

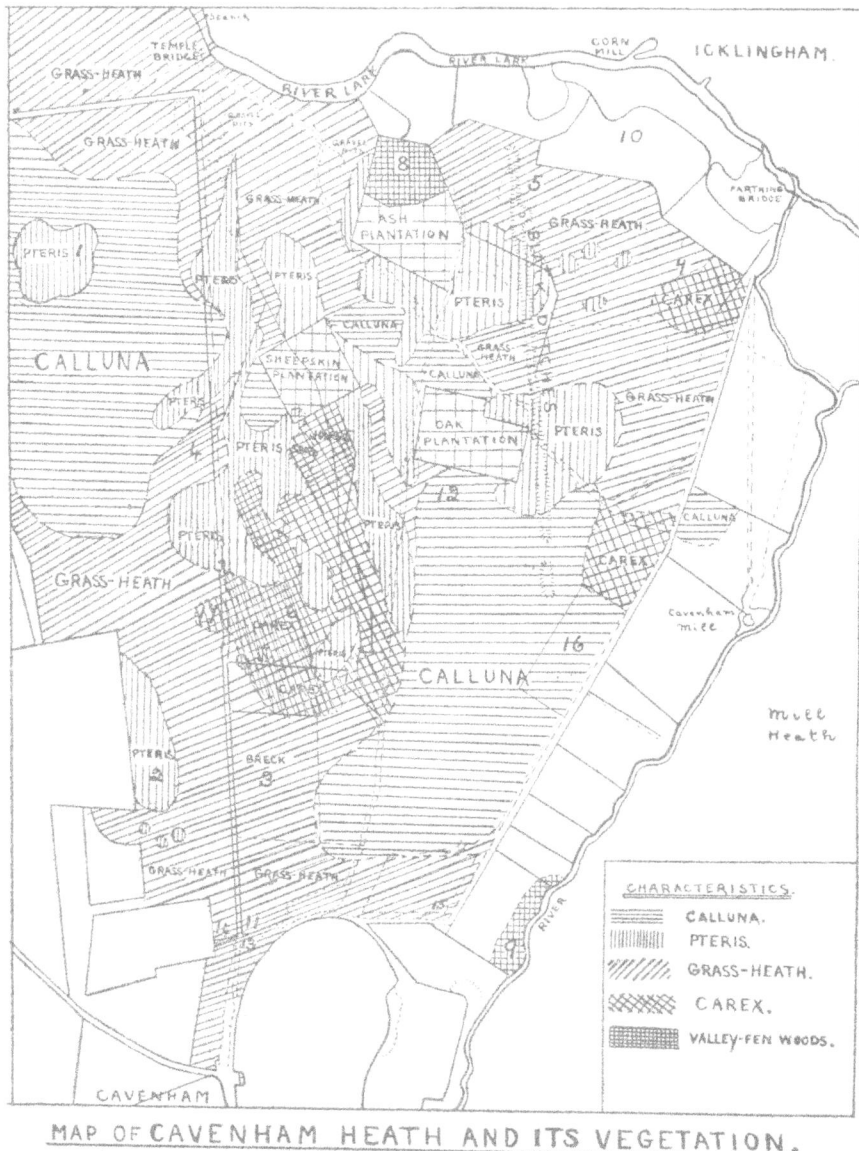

FIG. 3. For further explanation see text.

it will be seen that the areas of bracken are usually connected with the plantations, the areas marked 1 and 2 being exceptions; there are several small heathland meres near the regions marked 4 and 5; the areas of *Carex arenaria* are typically near valleys, as shown at the parts marked 6 and 7; valley-fen woods occur on the areas marked 8 and 9; and a single row of pines runs from the spot marked 11 to near the area marked 4. The areas marked 3 and 12 are typical "Brecks."

The Heath is inhabited by innumerable rabbits (some of the burrows appear in Photo. 3) which, since their natural enemies (stoats, etc.) have been kept down, have increased to such an extent as to exert great influence on the vegetation, a point which will be dealt with in some detail later.

Diggings were made in order to obtain sections of the soil occupied by the various types of vegetation; a typical section is shown in Fig. 4. (1) Among luxuriant *Calluna*, in the typical heather areas, the surface layer to a depth of about 5 cms. consists of raw humus made up of decaying *Calluna* leaves and fragments of *Cladonia* and various mosses. (2) A layer of humus and peat, very tough owing to the presence of heather roots; the lower limit of this layer is indefinite, since the main roots run down locally into the next layer, but its thickness varies between 4 and 8 cms. (3) A layer of dark peaty material mixed with sand; rather tough owing to the presence of *Calluna* roots, and usually about 8 cms. thick. (4) A very characteristic layer of a dark chocolate colour, consisting of sand darkened by humous compounds; thickness about 15 cms. (5) A layer of sand particles cemented by humous compounds; not usually hard but quite definite and constituting a moor-pan, a sort of soft black sandstone, usually about 5 cms. thick. (6) A stratum of sand particles cemented by iron compounds, forming an iron-pan usually about 7 cms. thick but often sending down cones to a distance of 25 cms. into the next layer (7) which consists of dark red sand and is usually about 30 cms. thick. (8) A stratum of light reddish sand about 30 cms. thick; below this the sand gradually gets lighter in colour as the depth increases.

A comparison with typical borings given by Graebner for the North German heaths shows that the characteristic soil layers of the East Anglian heaths correspond fairly well with those of the heaths of the North German Plain. Cavenham Heath, possessing as it does only slight pans and a very moderate production of raw humus, appears to be in these respects intermediate between dune-heath from which hard-pan is absent and raw humus scarcely developed and the typical *Calluna* heaths of various other parts of England which are associated with hard pan and abundant production of raw humus. The soil layers near the surface of these East Anglian heaths often contain appreciable quantities of lime; several determinations showed that in the soils occupied by *Calluna* the percentage of carbonates was from 1 to 2.

The results of three typical mechanical analyses of these soil layers are shown in the accompanying table; the large proportion of coarse sand in all cases will be noticed. *A* is an analysis of sand associated with flints

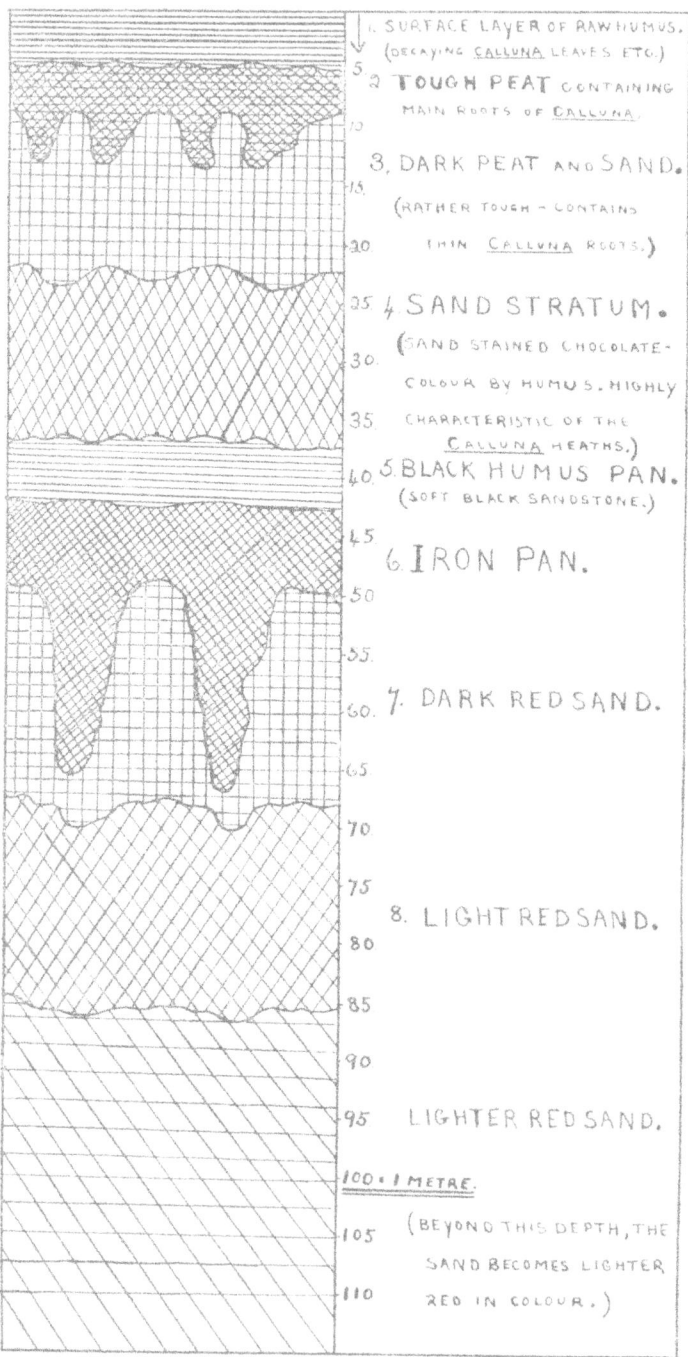

FIG. 4. Diagram of typical section through the upper strata of Cavenham Heath. The figures denote depths in centimetres.

underlying grass heath; *B* of roots and sandy humus from grass-heath; *C* of sandy humus from luxuriant *Calluna* heath. The numbers indicate percentages.

	A	*B*	*C*
Moisture determined at 100°C.	1·92	10·02	6·6
Humus	1·15	4·6	3·4
Mechanical analysis of mineral portion:			
Fine gravel (over 1 mm.) ...	1·24	3·54	1·34
Coarse sand (1—0·2 mm.) ...	78·00	66·00	63·00
Fine sand	13·60	20·18	15·50
Coarse silt	2·48	6·70	2·45
Fine silt	1·30	1·52	1·67
Clay	0·00	0·00	3·30

IV. Plant Associations on Cavenham Heath.

The principal vegetation units on Cavenham Heath are the following: (*a*) associations dominated by *Calluna vulgaris*; (*b*) grass-heath associations; (*c*) associations dominated by *Carex arenaria*; (*d*) associations of *Pteris aquilina*; (*e*) vegetation of the small meres; (*f*) valley-fen wood associations. The general positions of these are indicated on the map of the Heath (Fig. 3).

(*a*) **Calluna vulgaris Associations.** The associations dominated by heather usually tend to occupy the central portions of the upper areas. In the centre of these areas the bushes are typically close together and little else can grow except *Cladonia* spp. and a few mosses under the bushes, but towards the edges of the *Calluna* areas the bushes are further apart and the flora is much richer, various plants growing in the intervening spaces, while where the heather areas border those dominated by bracken, scattered bracken plants often occur among the heather. According to the gamekeepers the duration of life of the individual *Calluna* bushes in the district is about 15 years. Various patches of the heather appear to be dying out, and though this may be simply due in many cases to the plants having nearly reached the end of their natural life this general statement may often have to be modified owing to the existence of various immediate causes of death. There are many rabbit burrows among the *Calluna*, often associated with a characteristic flora which is dealt with later. Small oaks and pines occur occasionally among the thick heather. In damp low-lying places among the luxuriant *Calluna*, where the ground water often reaches to near the surface and the peat is thicker than elsewhere, a characteristic sub-association occurs of which *Erica tetralix* is the chief constituent, though on Cavenham Heath *Erica tetralix* never grows on the higher and drier areas amongst the abundant *Calluna*. On some of the East Anglian heaths north of Breckland the heath vegetation consists almost entirely of *E. tetralix*, but these are much damper than Cavenham Heath and *Juncus squarrosus* often occurs on them.

Since this heath is so dry and the rainfall is much lower than that of the upland moors which are also often occupied by *Calluna*, some data were obtained regarding the distribution of the root system of *Calluna* in the drier soil and climate for comparison. No roots were found in the surface layer which consists of dry peat made up of dead *Calluna* leaves and fragments of *Polytrichum piliferum*, *Dicranum scoparium*, *Cladonia coccifera*, etc. Most of the main and branch heather roots ramify horizontally at 5 to 15 cms. from the surface, though some of the main roots penetrate to 40 cms.

Plants occurring in the areas dominated by Calluna.

(i) Among thick *Calluna*: Calluna vulgaris **d**, Pteris aquilina **la**, Crataegus monogyna **o**, Pinus sylvestris **o**, Quercus sessiliflora **o**, Betula alba **o**, Polytrichum piliferum, Hypnum schreberi, Leucobryum glaucum, Dicranum scoparium, Ceratodon purpureus, Cladonia coccifera, C. cervicornis, C. alcicornis, C. sylvatica, C. furcata, C. uncialis.

(ii) In places where the *Calluna* bushes are less crowded there also occur, in approximate order of frequency: Festuca ovina, Agrostis vulgaris, Galium saxatile, Stellaria media, Alchemilla vulgaris, Draba verna, Teesdalia nudicaulis, Rumex acetosella, Aira praecox, Luzula campestris, Veronica officinalis, Plantago coronopus, Campanula rotundifolia, Solanum nigrum, Genista anglica, Conium maculatum, Lotus corniculatus, Urtica urens, U. dioica, Myosotis collina, M. versicolor, Sagina apetala, Cardamine hirsuta, Viola riviniana, V. canina, Dactylis glomerata, Potentilla erecta.

(iii) Among the damp heath sub-associations which occupy the various lower areas in the *Calluna* the following also occur: Erica tetralix **d**, Calluna vulgaris **f**, Salix repens **ld**, Juncus squarrosus **f**, Molinia caerulea **f**, Drosera rotundifolia, Aulacomnium palustre, Sphagnum compactum.

(*b*) **Grass-heath Associations.** The vegetation of the grass-heath forms a short close turf consisting chiefly of *Festuca ovina* and *Agrostis vulgaris*, but frequently the grass plants are wide apart and numerous annual dicotyledonous plants are present, many of these being characteristic early-flowering ephemeral species which finish flowering by the middle of May.

When the hotter drier weather comes, usually about the middle of May, the grass-heath associations which previously were of a green colour quickly become almost brown; and when the cooler damper weather comes, usually about the middle of October, the brownish grass-heath associations quickly become green again. This cyclic change in the "seasonal aspects" of these grass-heath associations is very marked and striking.

For comparison with the somewhat similar ephemeral plants of deserts and sand dunes which possess very deep root systems in proportion to the height of the subaerial part of the plant, data were obtained as to the depth of the subterranean portions of some of the plants of the grass-heath association. The typical root system of the annual plants, e.g. *Alchemilla vulgaris*, consists of a tap root which gives out fine laterals and may penetrate to a depth of 30 cms. or more into the sand, though the shoot usually does not rise more than 2 cms. above the surface of the soil. Often however in places where sand has been recently deposited by wind the root systems of the grass-heath plants are adventitious ones developed on sub-

PLATE II

Photo. 3. Orientated view on Cavenham Heath. A very typical Breckland view. Dwarf grass-heath in foreground, with rabbit burrows ; *Pteris aquilina* towards the right (above) ; *Carex arenaria* in foreground on right (below) ; *Calluna* heath in distance. (See p. 5.)

Photo. 4. Orientated view on Cavenham Heath, taken exactly a year after Photo. 3 and from the same position. Hence a comparison of Photos. 3 and 4 will show various changes that have occurred during a year in the area photographed. (See p. 13.) It will be seen along with other things that the sharp bracken frontier characteristic of Breckland has advanced a considerable distance from right to left by rhizome growth and that the sand cupolas (see Chapter VI) characteristic of the Breckland grass-heath and seen in the lower left-hand portion and also in front of the stake have become considerably taller and better developed during the interval of one year.

terranean stems.   The heath flora includes characteristic dwarf forms of *Festuca ovina, Aira praecox, Cerastium semidecandrum* and *Veronica arvensis.* Frequently the leaves of these grass-heath plants are arranged in rosettes lying close to the surface where wind velocity is low and the air is damp through evaporation from the soil; in this respect, as well as in various others, the vegetation of the grass-heath resembles that of fixed sand dunes[1]. Ephemerals with rosette foliage are *Myosotis collina, Cerastium tetrandrum, C. semidecandrum* and *Erophila verna*; other plants with this habit are *Rumex acetosella* and *Plantago coronopus.*   An interesting exception to the general dwarfing of the vegetation, which rarely rises more than 2 or 3 cms. above the surface, is found immediately around the decaying bodies of dead rabbits which occasionally occur on the heath; here there is usually a zonation of the taller vegetation in the centrifugal order *Rumex acetosella, Agrostis vulgaris, Festuca ovina, Luzula campestris,* ordinary grass-heath.

On many of the grass-heaths of Breckland the rabbit burrows are associated with a very characteristic flora, which includes *Urtica urens, U. dioica, Conium maculatum, Solanum nigrum, Veronica officinalis, Galium saxatile, Stellaria media, Alchemilla vulgaris, Myosotis collina,* and (very rarely) *Pteris aquilina* and *Lastrea* sp.   Later in the year the burrows are also characterised by various ephemerals which have died down on the exposed grass-heaths (see next lists).   On some of the grass-heaths considerably north of Breckland near Sandringham which are much damper than the Breckland heaths, the rabbit burrows are not picked out nearly so definitely by the vegetation. Thus the characteristic flora of the rabbit burrows is possibly associated with differences in the available water supply, but the presence of such stinging and poisonous plants as *Urtica urens, U. dioica, Conium maculatum* and *Solanum nigrum* in and near the burrows is probably due chiefly to the rabbits ignoring these unattractive plants while eating their normally more successful competitors.

Plants occurring on the Grass-Heath Areas.   The grass-heath areas are much richer floristically than are the areas dominated by *Calluna.*   The dominant plants are: Festuca ovina (dominant) and Agrostis vulgaris (co-dominant).   The following ephemeral plants occur: Erophila (Draba) verna, Teesdalia nudicaulis, Cerastium semidecandrum, C. tetrandrum, C. arvense, Arenaria serpyllifolia, A. tenuifolia, Myosotis collina, Erodium cicutarium.   The following plants which occur are rare in England outside of Breckland and are possibly remnants of a former steppe flora (see Tansley, *Types of British Vegetation,* p. 97): Silene otites, S. conica, Muscari racemosum, Artemisia campestris, Carex ericetorum, Medicago falcata, M. sylvestris, Veronica verna, V. triphyllos.   The following plants occurring in Breckland are usually absent from sandy inland heaths: Carex arenaria, Phleum arenarium.

In addition to the above the following plants occur on the grass-heath areas of Breckland: Aira praecox, Festuca rubra, F. myuros, Thymus serpyllum, Sedum acre, Trifolium minus, T. arvense, T. filiforme, Viola arvensis, V. riviniana, Silene inflata, Lychnis alba,

---

[1] **Oliver, F. W.** and **Salisbury, E. J.**   "Topography and vegetation of Blakeney Point." Botanical Laboratory, University College, London, 1913.

Convolvulus arvensis, Geranium molle, Ornithopus perpusillus, Potentilla argentea, Lotus corniculatus, Anthyllis vulneraria, Ononis repens, Vicia angustifolia, V. lathyroides, Galium verum, G. saxatile, Plantago coronopus, Myosotis collina, M. versicolor, M. arvensis, Veronica officinalis, V. arvensis, V. serpyllifolia, V. hederifolia, Campanula rotundifolia, Reseda lutea, Hieracium pilosella, Taraxacum erythrospermum, Rumex acetosella, R. conglomeratus, Senecio jacobaea, S. viscosus, Saxifraga tridactylites, S. granulata, Sagina procumbens, Crepis virens, Filago minima, Tillaea (Crassula) muscosa, Calluna vulgaris (dwarfed), Ulex europaeus (locally dominant), Pteris aquilina (locally abundant), Sambucus nigra (occasional), Crataegus monogyna (occasional).

In addition to the above 70 species of phanerogams characteristic of the grass-heaths 80 other less characteristic and less frequent species have been found on them making a total of 150 species of flowering plants.

The following bryophytes and lichens occur on the grass-heath: Polytrichum piliferum, Dicranum scoparium, Campylopus flexuosus, Tortula muralis, Hypnum schreberi, Leucobryum glaucum, Cladonia rangiferina, C. coccifera and forma cornucopioides, C. cervicornis, C. alcicornis, C. fimbriata, C. tubaeformis, C. furcata, Cladina sylvatica, C. uncialis, Cetraria aculeata, Lecidea granulosa.

(c) **Carex arenaria Associations.** It has already been mentioned that *Carex arenaria*, which is characteristic of coastal sand dunes, occurs in an inland habitat in Breckland (the positions of the *Carex* associations on Cavenham Heath are indicated in Fig. 3), and that these *Carex* associations are typically associated with the valley sides—it may be noted that before the tide-gates were placed in the Great Ouse nearly 200 years ago, salt tides used to come up the river Lark, and in immediately post-glacial times when the Wash reached far inland, this river was probably a salt-creek.

The *Carex* typically forms dense associations and crowds out practically everything else, but there are usually many rabbit tracks amongst the thick *Carex*, and along these and in the rabbit burrows *Festuca ovina*, *Galium saxatile* and *Stellaria media* manage to survive. With these exceptions the *Carex* association is a pure one (see Photo. 6). The *Carex arenaria* in Breckland is not confined to sheltered areas as it seems to be at Blakeney. Seeds are often blown from the *Carex* associations on to the grass-heath associations, and these frequently grow and colonise the grass-heath with *Carex arenaria*. The *Carex* is also spreading rapidly from the pure associations into the grass-heath areas by means of vigorous rhizomes.

(d) **Pteris aquilina Associations.** The positions of the *Pteris* associations on Cavenham Heath are indicated on the map of the area (Fig. 3). The bracken fern here typically forms pure associations with comparatively sharp and distinct edges (see Photo. 3). In most cases the transition seemed to be much too sharply marked to be attributed to edaphic conditions, and diggings made on the edges revealed the fact that young portions of the ends of the rhizomes had penetrated beyond the main bracken associations into the grass-heath association. It was thus discovered that the bracken fern is advancing and spreading, and the sharp border simply represents the position which it has reached at any given time.

PLATE III

Photo. 5. Rabbit burrow flora, Lakenheath Warren. The burrows are clearly indicated by their different vegetation (*Urtica dioica, Conium maculatum, Solanum nigrum*, etc.) (see p. 11); the grasses of the grass-heath have a more tufted habit than at Cavenham

Photo 6. Mildenhall Warren. *Carex arenaria* association in foreground, except around rabbit burrows; *Calluna* heath in distance; pine plantation in background. (See p. 12 and also pp. 34–42.)

In order to determine the rate of spread at various points, a number of quadrats[1] were made on the bracken borders in 1913, and many orientated photographs have been taken at recorded times[2]. Owing to the late frosts in 1914, most of the early young fronds were destroyed, and recharting the quadrats in 1914 would not have given a fair indication of its average progress. Many of the young fronds which were destroyed were however in advance of the previous year's fronds, so there is no doubt that the bracken is spreading (compare Photos 3 and 4), and will probably dominate almost the whole of the area in time unless means are eventually taken to check it. It seems able to spread everywhere except down into some of the damper valleys. At these places the spreading of the bracken seems to be definitely stopped. It is not known at present why this is so, but in these places the water table is often close to the surface, and any advancing bracken rhizomes would often be surrounded by stagnant water for long periods and would probably rot off. The vegetative spread of the bracken is usually completely stopped by very slight ditches.

With regard to Photos 3 and 4, Photo. 4 was taken exactly a year after Photo. 3 and from the same position. Hence a comparison of Photos 3 and 4 will show various changes that have occurred during a year in the area photographed. It will be seen that the bracken has advanced a considerable distance from right to left, thus definitely confirming the previously well-established belief that the sharp *Pteris* frontiers characteristic of Breckland are not stationary but are advancing, the bracken spreading vegetatively by rhizome growth. The sand cupolas—to be dealt with later—characteristic of the Breckland grass-heath and seen in the lower left-hand portion and also in front of the stake have become considerably taller and better developed during the interval of one year. The *Carex arenaria* tufts seen in the lower right-hand corner are also larger than they were a year ago, showing that the plant tends to spread; but although the recent rapid spread of *Carex* in Breckland is chiefly due to the presence of rabbits and their differential action upon *Calluna* and *Carex* respectively, the *Carex* in the region shown in the picture would probably have spread much more rapidly at the expense of the grass-heath but for the heavy rabbit pressure in this region which is partly caused by the recent spread of the unpalatable neighbouring *Pteris*.

[1] Becker's "anglemeter" has proved very useful for the rapid setting-out of quadrats and for subsequently picking up their positions on the extensive open heaths—by using the fact that the angles in the same segment of a circle are equal. For the actual charting in the field, a board arranged with rubber bands for the paper and fitted with a compass, like a cavalry sketching board but larger, has proved very useful. Squared paper already backed with linen is obtainable and is very durable and does not readily tear; and it is much more suitable than ordinary paper for charting in the field, which often has to be done in rough weather.

[2] See **Farrow, E. P.** "On a photographic method of recording developmental phases of vegetation." *Journ. Ecol.* **3**, 1915, p. 121.

The areas dominated by bracken are usually and typically associated with the artificial tree plantations on the heath, the areas marked 1 and 2 being exceptions. It seems probable that the bracken may have been introduced when the trees were planted over 100 years ago. As the bracken does not spread across narrow ditches reproduction by spores apparently does not typically occur here, but cases have been observed in which bracken is commencing to grow in rabbit burrows a long way from a main bracken association (see Photos 7 and 8). Here it has in all probability spread by spores; the bottoms of the rabbit burrows were apparently damp enough for the production of prothalli, although the surface of the grass-heath is far too dry for this to occur.

In Photo. 8 which shows a more advanced condition than Photo. 7 the *Pteris* has begun to spread vegetatively, producing a small island of bracken fronds far from a main bracken association. This suggests the view that if the *Pteris* was introduced with the pines and has normally spread vegetatively from the plantations the larger islands of bracken fronds (marked 1 and 2 in map of Cavenham Heath, Fig. 3) which are isolated from the plantations may have begun to grow in these positions owing to spore dissemination and prothallus production in rabbit burrows.

(*e*) **The Vegetation of the small meres.** It has already been mentioned that small specimens of the heathland meres which are characteristic of Breckland occur on Cavenham Heath, chiefly near the areas marked 4 and 5 on the map (Fig. 3). These meres (Photo. 9) are typically small circular pools occupying crater-shaped hollows. The current theory to account for them is that they occur over pipes in the chalk and that they may still be in process of formation. If this be so, the lower damper areas of the wet-heath sub-association—which, as already stated, occasionally occur amongst luxuriant *Calluna* heath and are chiefly characterised by *Erica tetralix*—may possibly be incipient meres. The water-levels in the meres vary greatly at different times, and doubtless this variation exerts a very great influence on the zonation and distribution of the characteristic vegetation. If the current theory of the formation of these small meres be true, the conditions of changes in the succession of their vegetation over long periods of time would be very interesting; as, if the meres gradually become deeper, the changes in the conditions would be quite opposite to the changes around most lakes, for, around most lakes, as is well known, the succession is towards the formation of dry land, instead of being towards an aquatic formation.

The zonation of the vegetation in the meres and in the surrounding heath moors is as follows. (1) In the water there grow *Glyceria fluitans* and *Ranunculus peltatus*. (2) Round the edge of the water there is usually a zone of *Juncus obtusiflorus*; this is typically a fen plant, and in this case it probably roots down through the acid surface layers into the subsoil.

PLATE IV

Photo. 7. *Pteris aquilina* growing in bottom of a rabbit burrow in grass-heath association. The damper conditions in the burrow were suitable for the production of prothalli, while the surface of the grass-heath is far too dry for this to occur. (See p. 14.)

Photo. 8. Isolated group of *Pteris* fronds spreading from a rabbit burrow. This photograph shows a more advanced condition than Photo. 7, for here the *Pteris*, which in all probability reached this position by spore dissemination and commenced to grow by prothallus production under the damper conditions obtaining in a rabbit burrow, has begun to spread vegetatively, producing a small island of bracken fronds far from a main bracken association. (See p. 14.)

PLATE V

Photo. 9. Small mere on Cavenham Heath, near area marked 4 in map (Fig. 3 in text). Heath-moor (*Erica tetralix, Molinia caerulea,* etc.) around the mere; *Ranunculus aquatilis* in the water; *Salix repens* on the right; *Pteris aquilina* on the right (distance); *Calluna vulgaris* (horizon). (See p. 14.) The water level in these meres varies greatly, and many of the *Erica tetralix* bushes immediately around this mere have been killed owing to long-continued sub-mergence during a protracted period of high water-level in 1912.

Photo. 10. Valley-fen wood in valley of the river Lark: *Alnus rotundifolia, Fraxinus excelsior, Betula alba.* Undergrowth of *Lastrea thelypteris, Phragmites vulgaris,* etc.

(3) Immediately outside (2) comes a zone characterised by *Juncus supinus*, succeeded by a zone of *Hydrocotyle vulgaris, Peplis portula* and *Aulacomnium palustre*. It is the last named zone which is chiefly affected by variations in the water level, and probably these variations largely determine the distribution of this zone. Next to this zone *Salix repens* frequently occurs; the *Salix* is apparently in many cases encroaching upon zones (3) and (4). Immediately outside the *Salix repens* and the *Juncus supinus-Hydrocotyle* zone, there is typically a broad zone (5) dominated by *Erica tetralix* and *Molinia caerulea*; it is this zone which is characteristic of the damp heath sub-association in the small lower areas amongst vigorous *Calluna*, and if these are really incipient small meres, the other more central aquatic zones have not yet appeared. On the drier upper edges of zone (5), scattered plants of *Calluna vulgaris* appear and the sub-association gradually passes into the main *Calluna vulgaris* association on the drier upper areas, or often into the bracken association, but the bracken never spreads into the damper areas.

Plants occurring in and around the small meres. In addition to the plants that are numerous enough to characterise definite zones, the following other plants occur around these meres: Juncus squarrosus, J. conglomeratus, J. acutiflorus, J. glaucus, J. effusus, Menyanthes trifoliata, Mentha aquatica, Agrostis vulgaris, Agrostis alba, Betula alba (seedlings), Potentilla palustris, P. erecta, Galium uliginosum, Lotus uliginosus, Radiola linoides, Scrophularia aquatica, Anagallis tenella, Peplis portula, Equisetum spp., Sphagnum spp.

(*f*) **Valley-fen Woods.** Fenland woods occur in the river valleys on the edges of Cavenham Heath on the areas marked 8 and 9 on the map (Fig. 3). Some artificial plantations have been made on the upper parts of the heath, two consisting entirely of *Pinus sylvestris* and one chiefly of *Quercus sessiliflora*; and there is also a small plantation of *Betula alba* and *B. pubescens*. Fifty years ago there were only a few pines and birches in the slight valley that forms the western border of the Heath, but now there are hundreds of them, hence it appears that in some cases pines and birches can colonise the valley sides from the bottom of the valleys. At the present time slight depressions of the general level on the edge of the heath are often picked out by pine and birch seedlings, probably because there is here less sand between the surface layers and the chalk, so the seedlings can obtain water more easily by capillarity from the chalk.

Near the river valleys, however, the actual water level is often near the surface, and luxuriant tree growth is possible owing to these special edaphic conditions. Ash and alder are highly characteristic of these valley-fen woods (Photo. 10); poplars (*Populus canescens, P. nigra* and *P. serotina*) also occur, besides willows (*Salix alba, S. fragilis, S. triandra, S. viminalis* and various hybrids), *Prunus spinosa, Crataegus oxyacanthoides, Ulmus glabra, Ligustrum vulgare, Betula alba, B. pubescens* and *Corylus avellana*. These valley-fen woods resemble in many respects the fen woods of the Norfolk carrs; ash and alder are very characteristic of both. The birches in these woods are usually *B. alba*, but if they are natural one would have expected them to have been *B. pubescens*. Sometimes areas bare of trees occur in the woods (Photo. 11); these open areas may be due to pasturing. These peaty edaphic woods often reach up on the higher portions to a wood containing *Quercus sessiliflora* (Photo. 12), in this respect resembling somewhat similar woods in Kent and Norfolk.

The trees are not sufficiently close together to interfere greatly with the undergrowth, which is luxuriant and is essentially that of fen, though in some places it approaches that of the glacial-clay woods, tending to resemble the ground flora of the damper parts of Gamlingay wood in Cambridgeshire[1], and presenting a striking contrast to the dwarf grass-heath vegetation with its ephemerals on the neighbouring upper dry sandy steppe-like areas.

Plants occurring in the Edaphic Woods. Alnus rotundifolia, Fraxinus excelsior, Betula alba, B. pubescens, Salix alba, S. triandra, S. purpurea, S. fragilis, S. viminalis (and various hybrids—alba × fragilis, etc.), Populus nigra, P. italica, P. canescens, P. serotina, Prunus spinosa, Crataegus oxyacanthoides, Ulmus glabra, Corylus avellana, Ligustrum vulgare, Geum rivale, Spiraea ulmaria, Urtica dioica, Agrostis alba, Listera ovata, Solanum dulcamara, Scabiosa succisa, Aira caespitosa, Helianthemum chamaecistus, Genista anglica, Lysimachia vulgaris, Lythrum salicaria, Carex disticha, C. paniculata, C. hirta, C. riparia, Ajuga reptans, Nepeta hederacea, Phragmites vulgaris, Epilobium palustre, Habenaria conopsea, Lychnis flos-cuculi, Mentha aquatica, Potentilla palustris, P. anserina, Juncus obtusiflorus, J. conglomeratus, J. acutiflorus, Galium uliginosum, G. aparine, Lotus uliginosus, Myosotis palustris, Polygonum amphibium, Malva sylvestris, Calamagrostis canescens, Veronica Beccabunga, V. officinalis, Menyanthes trifoliata, Lycopus europaeus Scutellaria galericulata, Lastrea thelypteris, L. filix-mas, L. aristata, Polystichum aculeatum Pteris aquilina, Equisetum spp.

[1] **Adamson, R. S.** An ecological study of a Cambridgeshire woodland." *Journ. Linn. Soc.*, **40**, 1912

PLATE VI

Photo. 11. Valley-fen wood, valley of the Lark: *Alnus rotundifolia, Populus nigra, Salix* spp., *Fraxinus excelsior*, undergrowth of *Spiraea ulmaria, Geum rivale, Urtica dioica*, etc.

Photo. 12. Valley fen-wood, valley of the Lark: *Quercus sessiliflora, Betula alba, Pinus sylvestris*; undergrowth of *Calluna vulgaris, Carex arenaria, Agrostis vulgaris, Rubus fruticosus, Galium saxatile, Hydrocotyle vulgaris*.

PLATE VII

Photo. 13. Bare area and dwarfed *Calluna* bushes around rabbit burrows in *Calluna*-heath. (See p. 18.) *Cladonia* spp. amongst bushes in foreground (shows whitish in photograph).

Photo. 14. Rabbit-proof cage on degenerating *Calluna* zone, to show difference between the *Calluna* plants inside and outside the cage. The *Calluna* bushes inside the cage possess many leaves and flowers (show whitish), whilst outside the cage the *Calluna* plants are leafless and degenerating—note scraggy leafless bush on the right of the cage and degenerating *Calluna* bushes in the distance. (See p. 18.)

CHAPTER II.

FACTORS RELATING TO THE RELATIVE DISTRIBUTIONS OF *CALLUNA*-HEATH AND GRASS-HEATH IN BRECKLAND[1].

SYNOPSIS.

As has been stated in Chapter I, *Calluna*-Heath alternates with grass-heath in Breckland and detailed descriptions of these two characteristic associations have been given with a map of their distribution on Cavenham Heath (see p. 6). One of the first problems which arose during the course of this Breckland work was to determine the causes of the relative distributions of these two main associations. This alternation of grass-heath with *Calluna*-heath in Breckland has previously been supposed to be due to varying proportions of lime in the soil[1]. Examination of the upper strata of the soils of the grass-heath associations showed that they usually possess the dark chocolate coloured stratum at 20 to 35 cm. from the surface which is very characteristic of the *Calluna* associations.

FORMER DISTRIBUTION OF THE TWO ASSOCIATIONS.

During the examinations of the upper strata of the soils of the grass-heath associations on Cavenham Heath, some rather decayed remains of *Calluna* roots were found even near the outer edge of the grass-heath association, near the area marked 14 on the map of Cavenham Heath[2], about half a mile from the main *Calluna* association. This was a surprising discovery, for it meant that *Calluna*-heath must once have existed on this spot which is now grass-heath, and that probably the *Calluna* existed in this position fairly recently, since organic bodies apparently decay quickly in this open sandy soil. A closer examination was therefore made of the transition zone where the grass-heath association changes into *Calluna*-heath. In this zone, as the typical *Calluna* association is approached from the grass-heath association, isolated plants of *Calluna* of small height begin

---

[1] "Types of British Vegetation," 1911, p. 107.   [2] See p. 6.

F

2

to appear, and these gradually become more numerous and of greater height as the typical *Calluna* association is entered. This transition zone is also characterised by the presence of abundant *Cladonia coccifera, C. cervicornis* and *C. alcicornis* growing amongst the branches of the more dwarf *Calluna* plants (Plate VIII, Photos 15 and 16).

EXPLANATION OF THIS FORMER DISTRIBUTION.

These phenomena were very puzzling for some time—the luxuriant *Cladonia* appeared to be smothering the *Calluna*. The clue to the matter was however given by an examination of the areas immediately around rabbit burrows which occur amongst the *Calluna* plants on the heath, as well as in the places occupied by various other associations. Immediately near the rabbit burrows the ground is often bare or is occupied only by dwarfed grasses. Just around these areas the *Calluna* bushes are much less in height than typical ones and have a smooth rounded appearance with no projecting shoots, and they become taller further away from the rabbit burrows. The small *Calluna* bushes in this zone around rabbit burrows in the *Calluna*-heaths resemble in many respects the small bushes in the transition zone between *Calluna*-heath and grass-heath. From these phenomena it thus appeared that the main *Calluna*-heath associations might possibly be degenerating to grass-heath through rabbit attack.

It was thus thought advisable to examine the rabbit dung in the laboratory for traces of fragments of *Calluna* leaves, and with this object in view, fragments of rabbit excrement from the transition zone were broken up and boiled alternately in dilute acid and in dilute alkali and filtered off between the boilings, in order to separate out the fibrous portions. The particles which remained after this process were then compared with particles obtained by treating chopped-up *Calluna* leaves in the same way, and after microscopic comparison of the two sets of particles there was no doubt that the rabbit dung contained fragmentary remains of *Calluna* leaves.

Thus it appeared that the rabbits certainly eat the *Calluna* leaves, but whether they eat them to a sufficient extent to cause the main *Calluna*-heath associations to degenerate to grass-heath, except possibly just around the rabbit burrows, remained uncertain. A rabbit-proof cage was erected in the middle of the transition zone (see Plate VII, photo. 14), and a great difference between the *Calluna* inside and outside the cage quickly became apparent; the *Calluna* inside the cage recovering from the previous rabbit attack and producing many fresh leaves, while the *Calluna* outside the cage continued to degenerate.

Hence there is no doubt in this case about *the degeneration of* Calluna-*heath to grass-heath through rabbit attack*. In order to obtain further information about the rapidity of the degeneration, a belt-transect 150 metres long and one metre wide has been staked out across a typical area of the degenerating zone, and various means have been adopted for recording

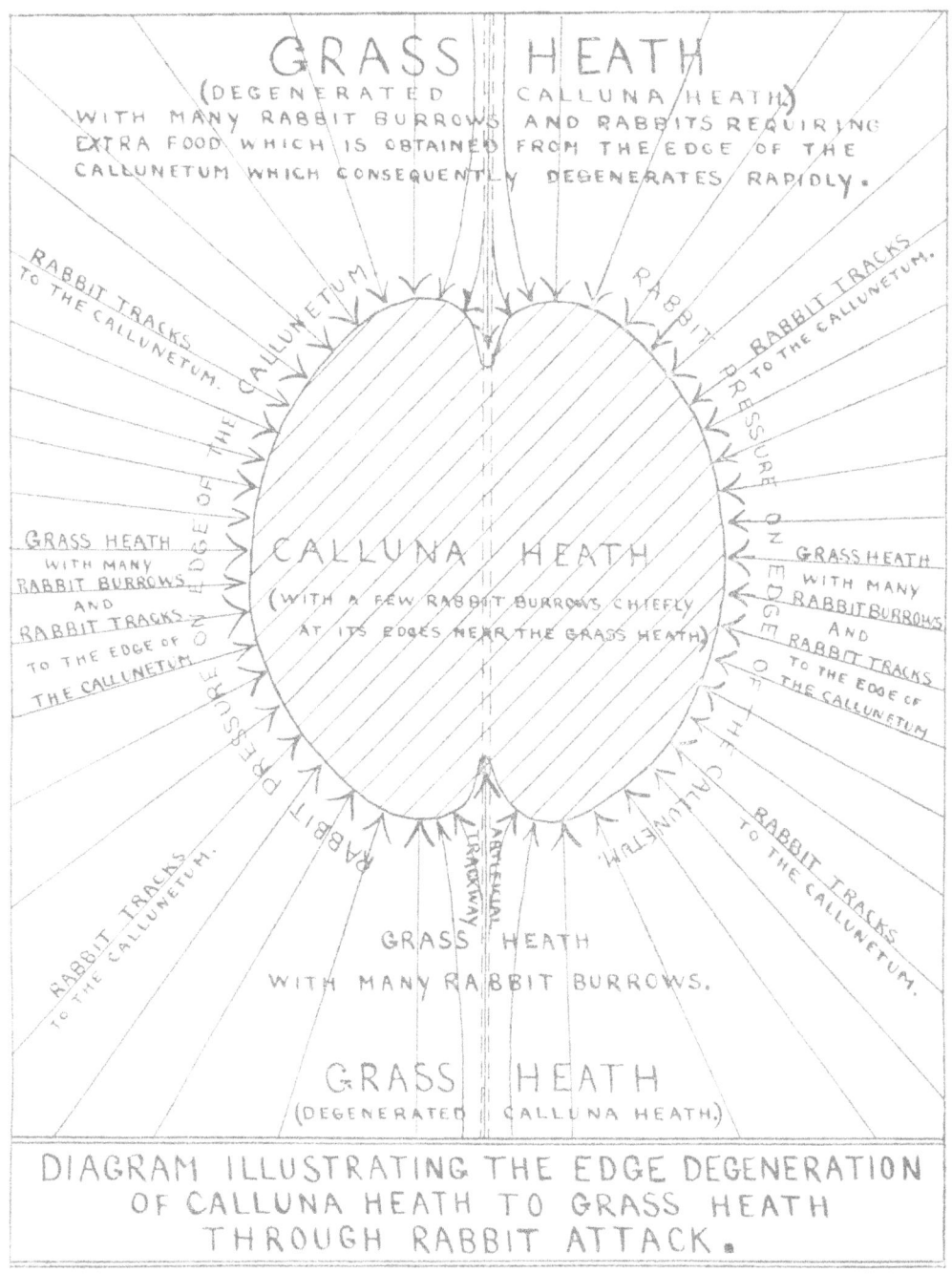

FIG. 5. Degeneration of *Calluna*-Heath. For further explanation see Text.

the temporary conditions of the *Calluna* bushes at fixed places along this transect at certain dates.

One of the methods has been to take various orientated photographs[1] at fixed places at intervals (in this case every 20 metres) along the transect. Photos 15 and 16 (Plate VIII) are two of the photographs taken, 20 metres apart, in this way along this transect across the degenerating zone. This long transect across the degenerating zone has also been very carefully charted, the sizes and heights of all the bushes being recorded. The detailed distribution of *Cladonia* along the transect has been recorded. This chart is on a scale of $\frac{1}{10}$ and is thus 14 metres long and it shows these phenomena in considerable detail. A second transect has also been made across a zone in which the *Calluna* is degenerating very rapidly, and orientated photographs have been taken along this second transect for the purpose of recording the condition of the bushes at certain positions at certain times. This shorter transect has also been charted on a similar scale to the longer one, and the sizes of the bushes and the distribution of the *Cladonia* on this transect have been recorded. It is intended to re-chart these transects sometime in the future, in order to see the details of what has happened in the meantime, and the transects will be re-photographed in exactly the same manner in which they were photographed before, in order that information relating to the details of the degeneration may be derived by comparison of the respective sets of photographs.

It was suggested that a good indication of the amount of rabbit attack to which the vegetation in various places had been subjected might be given by the relative numbers of pellets of rabbit dung per unit area in those places—as this index would tend to integrate the number of rabbits and the time they were present on any given area.

It was realised however that if the integration of the total rabbit attack was to be made over fairly long periods of time a correction might have to be applied to these numbers in order to allow for the decomposition and disintegration of some of the individual dung pellets in those areas which had been occupied by rabbits for a long time. Nevertheless the respective numbers of pellets per unit area might be a good measure of the total rabbit attack which had occurred at any spot for some time previously.

Unfortunately in fairly exposed positions the rabbit dung pellets are often somewhat blown about by wind and, owing to this, it was only possible to obtain an approximate accuracy of the numbers over fairly large areas, and it was necessary to take small sample areas and estimate the total numbers statistically. But with these precautions comparative numbers were obtained. It was found that the average numbers of rabbit dung pellets per square metre were: at the *Calluna*-heath end of the long transect,

[1] **Farrow, E. P.,** "On a Photographic Method of Recording Developmental Phases of Vegetation," *Journal of Ecology*, **3**, 1915.

PLATE VIII

Photo. 15. Orientated photograph on long transect across the transition zone between *Calluna*-heath and grass-heath where the *Calluna*-heath is degenerating to grass-heath owing to rabbit attack. (See p. 20.) Note rounded shape of the rabbit-grazed *Calluna* bushes.

Photo. 16. Orientated photograph on the same long transect 20 metres nearer the grass-heath than Photo. 15. The degeneration of the *Calluna*-heath to grass-heath has here proceeded considerably further than in the previous photo. Note the more dwarfed character of the rabbit-grazed *Calluna* bushes. Much *Cladonia* is seen growing amongst the eaten-down *Calluna* branches (shows whitish in the photo). (See p. 21.) Rabbit tracks are seen in the foreground of the picture.

120 ± 23; at the middle, 150 ± 37; at the grass-heath, 190 ± 46. These numbers may be taken as representing approximately the relative severities of the total rabbit attack at the various places along the transect during a certain period of time. It will be noticed that the total rabbit attack has been greater where the degeneration of the *Calluna* bushes has been greater.

DETAILS OF THE BIOTIC DEGENERATION.

The chief points about the process of degeneration of *Calluna*-heath to grass-heath owing to rabbit attack are as follows. The attacking rabbits invade the *Calluna* association between the bushes, eventually forming established tracks. They nibble at the *Calluna* leaves which are within their reach along the sides of these tracks, and those parts of the bushes which are within reach of the rabbits take on a characteristic rounded form (Plate IX, Photo. 18). Often, however, the central portions of the bushes are out of reach of the rabbits, and these parts remain tall while only the lower outer portions are rounded off.

The result of this attack is that the once closed *Calluna* association becomes more open in character, and various grasses (chiefly *Agrostis vulgaris* and *Festuca ovina*)—the forerunners of the grass-heath association—appear in the rabbit tracks along with various other plants. At the same time *Cladonia coccifera, C. cervicornis,* etc., *Leucobryum glaucum, Hypnum schreberi, Dicranum scoparium,* etc., which previously formed a subordinate layer of vegetation under and around the once luxuriant *Calluna* bushes, take on a new and enlarged lease of life. The *Cladonia* especially becomes active and grows vigorously amongst the eaten-down branches of *Calluna* (Plate VIII, Photo. 16). As the branches of the *Calluna* bushes are eaten down more and more, the *Cladonia* becomes more and more luxuriant, apparently finding a very suitable habitat amongst the eaten-down branches of *Calluna*. The masses of eaten-down *Calluna* branches with *Cladonia* growing densely amongst them retain water very well—like sponges—and are usually very damp during the autumn and winter. Apparently largely owing to this dampness and smothering by the *Cladonia*, the eaten-down *Calluna* branches eventually begin to decay (they are badly decaying in the region shown in Photo. 16), disintegrate, break off at the surface and with the associated *Cladonia* they are eventually blown away, leaving behind the grass-heath association, which has been gradually expanding from the rabbit tracks.

Although the original rabbit-eaten *Calluna* branches which are above the surface of the soil and associated with the damp *Cladonia* decay fairly rapidly, break off at the surface and are blown away along with the associated *Cladonia*; yet the roots of the *Calluna* plants, which are beneath the surface of the soil, and protected from the above mentioned disintegrating agencies, often live on for some time after the destruction of the original subaerial portions, and these protected *Calluna* roots often endeavour to produce fresh subaerial branches. The rabbits however keep these fresh young *Calluna* stems nibbled down close to the surface and in the normal case the protected *Calluna* roots below the surface eventually die and decay.

When first these phenomena were seen, and before their significance was realised, it appeared that the luxuriant *Cladonia* was gradually smothering and killing off the *Calluna* bushes; but apparently the *Cladonia* only does this when associated with rabbit attack on the *Calluna* bushes, and during damp weather in the autumn and winter It is interesting to note that when the degenerating *Calluna* and the associated luxuriant and formerly smothering *Cladonia* inside the rabbit-proof cage on the transition zone (Photo. 14) were cut off from rabbit attack by the erection of the cage, the *Calluna* in its turn has been able to smother and kill the *Cladonia* which was formerly smothering it.

Often when rabbits attack old *Calluna* bushes, all the lower leaves on the taller branches are eaten off, leaving small and densely crowded clumps of leaves on the extreme ends of the branches; possibly these densely aggregated terminal clumps of leaves give the maximum assimilatory surface while exposing the least leaf surface to the rabbit attack. Only in comparatively few cases is old age the immediate cause of death of the *Calluna* bushes on Cavenham Heath. In the majority of cases the immediate cause of death is rabbit attack either alone or combined with the smothering effect of *Cladonia*.

OTHER FACTORS IN THE BIOTIC DEGENERATION.

When the branches of the *Calluna* bushes have been eaten down to a certain extent by rabbits, various other plants besides *Cladonia* often grow amongst these eaten-down *Calluna* branches. *Leucobryum glaucum* is one of the commonest and most deadly of these other injurious associates of the eaten-down branches of rabbit-attacked *Calluna* (Plate IX, Photo. 17). The *Leucobryum* typically forms a very dense growth, almost like a solid mass of tissue (see Photos 17 and 18), enclosing many of the eaten-down *Calluna* branches, which are smothered and die and eventually rot away. The writer has also seen *Cladonia* and *Leucobryum* crowding out rabbit-attacked *Calluna* on the north-west Norfolk heaths and also sheep-eaten *Calluna* hummocks on various mountains, and this appears to be a widespread phenomenon. Occasionally other mosses, such as *Dicranum scoparium* and *Hypnum schreberi*, also grow in rabbit-eaten *Calluna* bushes, but these do not usually form dense masses and they do not usually kill parts of the *Calluna* bushes.

The writer has observed the degeneration of *Calluna*-heath to grass-heath primarily owing to rabbit attack at many places in Breckland and it is a widespread phenomenon in this part of England.

PLATE IX

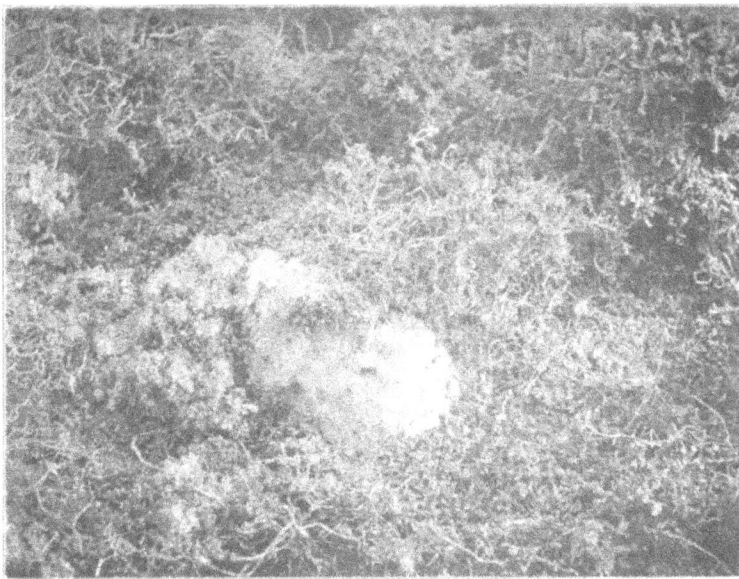

Photo. 17. Rabbit-attacked *Calluna* bush with the eaten-down *Calluna* branches smothered by a dense growth of *Leucobryum glaucum* (in the centre) and by *Cladoni coccifera* (on the left). The dense growths of *Leucobryum* and *Cladonia* amongst and above the eaten-down *Calluna* branches smother them and cause them to decay. (See p. 21.)

Photo. 18. Rabbit-grazed and rounded *Calluna* plant with many of the eaten-down branches badly crowded out and smothered by a dense growth of *Leucobryum glaucum*. (See p. 22.) Rabbit dung pellets and fragments of disintegrated *Calluna* bushes are seen in the foreground on the left.

EXPLANATION OF THE EDGE DEGENERATION OF THE *CALLUNA* HEATH.

In Breckland the degeneration from *Calluna*-heath to grass-heath usually takes place chiefly at the edges of the *Calluna* associations, whereas one might have expected the *Calluna*-heaths to degenerate uniformly all over. The probable explanation is as follows. At first the rabbits live chiefly upon the more luxuriant grass vegetation down the valley sides outside of the *Calluna* associations. Only comparatively few rabbit burrows occur near the centres of the *Calluna*-heaths a long way from the grass-heaths. Apparently the rabbits greatly prefer grass to heather as their main source of food supply. As the rabbits increase in numbers, however, and become relatively short of food in the valleys, some of them have to migrate from the valleys to the edges of the *Calluna* association on the higher ground for the balance of food supply which they require, and these rabbits tend to make burrows nearer to the edge of the *Calluna* association. There is thus a line of rabbit pressure along the edge of the *Calluna*-heath. If the rabbits are continually increasing in numbers, more and more rabbits are forced to go to the edge of the *Calluna*-heath for the balance of food supply which they require and the rabbit pressure on the edge of the *Calluna*-heath gradually increases and the *edge* of the *Calluna*-heath association gradually retreats. Later additional rabbits tend to make their burrows still further from the valleys near the new edge of the *Calluna* which they are forced to attack owing to the shortage of better food in the valleys. The occupants of the burrows near the *previous* edge of the retreating *Calluna* association also have to go to the *new* edge for their extra food, and the effect of these rabbits is added to that of the occupants of the new burrows near the new edge which thus—if the rabbits are increasing in numbers—continually tends to move more and more rapidly backwards. Also as the central *Calluna*-heath degenerates the total length of its edge decreases in length, and the length of the *Calluna* edge—and consequently the amount of extra food obtainable by a given penetration of the edge—on any given sector of the heath diminishes as the degenerating edge approaches the centre point. In consequence of this the rabbits have to penetrate the degenerating edge more and more in order to obtain a given amount of extra food from the edge and this helps to cause the degenerating edge to move backwards still more rapidly.

When the natural enemies of the rabbits are kept down, as they are at Cavenham, the rabbits eventually become extremely numerous and increase almost up to the limit of subsistence. The swarms of rabbits produce a great increase in the number of rabbit burrows on the degenerated grass-heath, and when as at Cavenham the rabbits on the grass-heath have ultimately been cut off from the more rapidly growing vegetation of the valleys by the erection of a rabbit-proof fence around the heath, they are forced to eat down the grass of the grass-heath very closely, and very many of them have to go to the edge of the *Calluna*-heath on the central upper portions for any necessary balance of food supply which they may require. The consequence of all this is that the rabbit pressure all along the line of the *edge* of the *Calluna*-heath at any time continually tends to increase, and the *Calluna*-heath degenerates at its *edge* with continually increasing rapidity.

These phenomena are illustrated in Fig. 5.

BEARING OF THE BIOTIC EDGE DEGENERATION OF *CALLUNA* HEATH UPON GRAEBNER's LEACHING HYPOTHESIS.

The fact that *Calluna*-heath usually degenerates sharply at its edges when the degeneration is caused by biotic attack, and the probable reason which is explained above of such edge degeneration, may also be interesting in view of the fact that many degeneration processes—such as that of oak-birch woodland to heath—usually occur chiefly at the edges of the degenerating association instead of uniformly throughout its area, even although the soil conditions may be fairly uniform throughout its area. In these cases a certain amount of biotic attack is usually present and probably the biotic attack acts upon, and produces the edge degeneration of, these other associations in the same way that it causes the edge degeneration of *Calluna*-heath, viz. the necessity of obtaining a balance of extra food from a different kind of plant. The analogy of the edge degeneration in these other classes of cases with the edge degeneration of *Calluna*-heath may rather tend to suggest that these other cases of edge degeneration are very likely *chiefly* due to some form or other of biotic attack upon the edge of the degenerating association acting in the same way as the biotic attack upon the *Calluna*, and not to such things as leaching of the soil, which if they exist are very likely merely secondary effects of the biotic degeneration.

There is a human trackway across part of Cavenham Heath and the *Calluna* has degenerated to a greater extent at the places where this trackway enters and leaves the *Calluna*-heath. (See Fig. 5.) Probably the chief cause of this is that the rabbits from the degenerated area can reach the luxuriant central *Calluna* for their balance of food supply more readily along this trackway than across the ordinary degenerating zone and that the *Calluna* bushes *en route* have incidentally become more eaten down.

EFFECT OF DIFFERENTIAL RATES OF INJURY UPON COMPETITION.

It is very interesting to note that on Cavenham Heath and elsewhere the rabbits very severely injure the grass-heath and keep it nibbled down very closely to the surface of the soil, and yet that they enormously benefit it, since if it were not for the rabbits the grass-heath would not exist at all, but would become replaced by heather. The grass-heath owes its very existence to an extremely injurious influence which nevertheless greatly benefits it because it injures its competitor slightly more.

CHAPTER III.

GENERAL EFFECTS OF RABBITS ON THE VEGETATION.

It was shown in Chapter II that rabbits produce so great an effect upon the vegetation of Cavenham Heath and other heaths of Breckland as to cause the main *Calluna* heath associations to degenerate to grass-heath and tc determine the relative distributions of these two main associations. It therefore seemed advisable to devote considerable attention to any other effects which rabbits might have upon the vegetation. It is known that rabbits exert a very great effect upon vegetation elsewhere in the country, and as a single instance the influence of rabbits on the vegetation of Blakeney Point in Norfolk may be cited[1].

RABBIT-PROOF FENCES.

In order to see what other effects rabbits might have upon the vegetation, the rabbit-proof fence which runs across the southern end of Cavenham Heath was first examined. This southern fence runs from the point marked 14 to the point marked 15 on the map on p. 6. At the time of examination it had been erected 14 years. There is a sharp change in the character of the vegetation between the two sides of this fence (see Pl. X, Photos 19 and 20). The difference chiefly manifests itself in the much greater height and luxuriance of the vegetation and in the far greater number of inflorescences on the

---

[1] **Oliver, F. W.** "Some Remarks on Blakeney Point, Norfolk." *Journal of Ecology*, **1**, No. 1, pp. 14 and 15.

protected side of the fence. On the unprotected side there are practically no flowers and much bare sand is exposed (see Photo. 19).

The area on the protected side of the fence is also far richer floristically than the larger unprotected area of Cavenham Heath, containing many plants which are not found at all on the main portion of the heath, which is subject to rabbit-attack. The following are some of the chief plants which occur on the protected side of the rabbit-proof fence whilst they are either rare or absent on the unprotected side: *Convolvulus arvensis, Silene inflata, Lychnis alba, Ornithogalum umbellatum, Lotus corniculatus, Geranium molle, Crepis capillaris,* and *Achillea millefolium.* Their absence on the unprotected side is probably chiefly due to the far greater amount of rabbit-attack in that position.

A comparison of Photos 19 and 20 shows that there is a greater difference in the vegetation on the two sides of the fence in Photo. 20 than in Photo. 19. On the protected side of the fence in the region shown in Photo. 20, *Carex arenaria, Urtica dioica* and *Agrostis vulgaris* are growing fairly luxuriantly— much more so than the vegetation on the protected side in Photo. 19. This is probably chiefly because the region shown in Photo. 20 is at a lower level and in consequence not so dry as that shown in Photo. 19; so that water supply is probably not so severe a limiting factor to the growth of the vegetation. Although the protected vegetation can thus grow taller in this region yet the vegetation on the rabbit-attacked side of the fence is eaten nearly as close as that shown in Photo. 19, the heavy rabbit-attack keeping it down nearly to the same low level in both cases. Nevertheless the rabbit-attacked vegetation in Photo. 20 forms a somewhat closer turf than that in Photo. 19 and not so much bare sand is exposed.

In the region shown in Photo. 20 it should be noted that *Carex arenaria* is largely dominant on the protected side of the fence, and is absent on the rabbit-attacked side. The heavy rabbit-pressure in this region is sufficient to prevent this plant from effectually invading the rabbit-attacked area by rhizome growth through the fence and thus eventually smothering the grass vegetation.

Carex arenaria is also present on the protected side of the rabbit-proof fence on the western side of the heath and is absent on the rabbit-attacked side. The same state of things also occurs in various other localities in Breckland.

PLATE X

Photo. 19. Rabbit-proof fence across the south side of Cavenham Heath--Western end. Note difference in the vegetation on the two sides. On the protected side there are many inflorescences, while on the unprotected side there are no inflorescences, the vegetation is nibbled closely down and much bare sand is exposed. (See p. 26 of the text.)

Photo. 20. Southern rabbit-proof fence on Cavenham Heath—Eastern end (looking in the opposite direction to Photo. 19). Note greater difference in the luxuriance of the vegetation on the two sides. Explanation of this greater difference is given in the text. Note presence of *Carex arenaria* on the protected side of the fence and its absence on the rabbit attacked side. (See p. 26.)

PLATE XI

Photo. 21. *Sedum acre* flowering inside the rabbit-proof cage on the relatively protected side of the fence. The *Sedum acre* flowers can be well seen in the photo. while outside the cage no flowers of this plant occur although the plant itself is freely present. There are also far more *Crepis, Carex, Agrostis, Festuca* and *Holcus* inflorescences inside than outside the cage and the vegetation is considerably more luxuriant as can be seen in the photo. (See p. 27.)

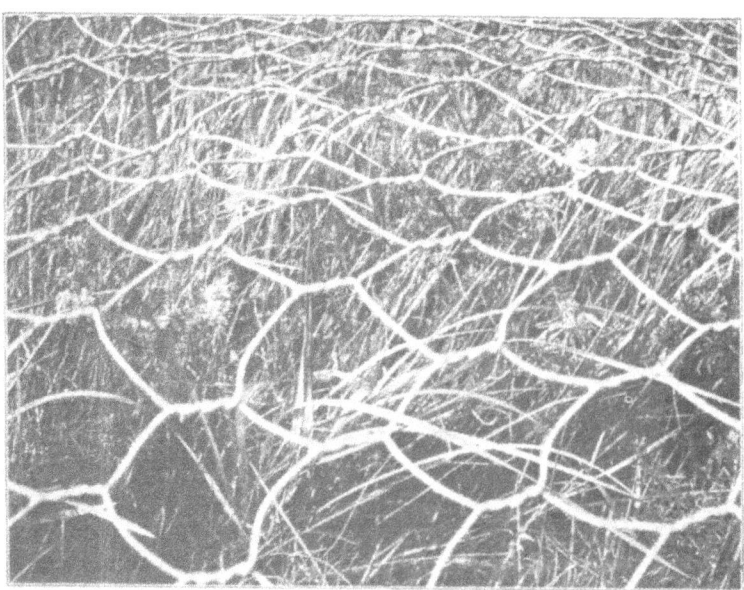

Photo. 22. *Galium verum* flowering inside rabbit-proof cage. *Galium verum* flowers inside the cage can be well seen in the photo. while outside the cage flowers of this plant are very scarce and those which do exist are very poor. (See p. 27.) Note also the luxuriance of the vegetative portions of the *Galium* inside the cage. Outside the cage the vegetative portions of this plant are eaten down very closely by the rabbits.

A few rabbits occur on the relatively protected side of the rabbit-proof fence seen in Pl. X, Photos 19 and 20. In order to determine the effect of this small number of rabbits on the vegetation, a rabbit-proof cage was erected on the protected side of the fence[1].

For some time there was only a small difference between the vegetation inside and outside the cage, but *Galium verum, Crepis capillaris* and *Sedum acre* grew and flowered appreciably better relatively to the grasses inside than outside the cage (see Pl. XI, Photos 21 and 22), showing that these plants suffered more than the grasses from the relatively few rabbits present on this side of the fence. After a period of three years the difference between the vegetation inside and outside the cage became much more marked, showing that the attack of even a few rabbits may have a considerable cumulative effect on the vegetation. *Rumex acetosella, Sedum acre, Galium verum, Crepis capillaris, Carex arenaria, Agrostis vulgaris, Festuca ovina* and *Holcus lanatus* eventually grew and flowered much better inside the cage than outside. Inflorescences of the four first named plants were either non-existent or very rare outside the cage although they occurred freely inside (see Pl. XI, Photos 21 and 22). It was very apparent that on the whole plants with tall vertical shoots and flowers eventually suffer more from increasing rabbit-attack than plants with shoots which can grow very close down to the surface of the soil. Probably vertically growing and tall shoots and inflorescences attract the attention of the rabbits, and are easier for them to cut off than more dwarf growing plants. Such plants in grassland are also usually fewer in individuals per unit area than the grasses and are thus more readily exterminated.

Sometimes ephemerals grow and flower when exposed to fairly heavy rabbit-attack and at first sight these might appear to be exceptions to the generalisation that vertically growing shoots and flowers suffer especially heavily from rabbit-attack; but the explanation in these cases probably is that though the ephemerals are as a whole heavily attacked by rabbits, their short annual period of vegetation exposes them to attack for a considerably less period than is the case with most other plants and that thus some of them manage to survive the rabbit-attack and are able to flower.

---

[1] By means of rabbit-proof cages the rabbit factor can be entirely removed from the environment and its separate effect studied, if allowance is made for time. It is rarely that we can thus completely isolate an ecological factor in order to study its special effects. Wire netting of one inch mesh should be used, since it is possible that very small rabbits can squeeze through $1\frac{1}{4}$ inch meshes. Netting of heavy gauge should be used for the sake of durability.

RABBIT-PROOF CAGE ON THE RABBIT-ATTACKED
GRASS-HEATH ASSOCIATION.

Another rabbit-proof cage (see Pl. XII, Photos 23 and 24) was fitted up on the unprotected side of the rabbit-proof fence in the middle of the degenerate grass-heath association on Cavenham Heath, about half way between the point marked 13 on the map and the present degenerating edge of the *Calluna* heath.   The rabbit-attack in the region where this second cage was erected is extremely heavy.   The portion of the degenerate grass-heath which was enclosed inside this second rabbit-proof cage quickly began to go back to *Calluna* heath (as can be well seen in the above photos).   It has already been mentioned that the *Calluna* roots remain alive in the soil for some time after all the upper portions of the *Calluna* plants have been killed by rabbit-attack and decayed by the damp luxuriant growth of *Cladonia*; and that the still living *Calluna* roots endeavour to send up fresh stems which are however normally quickly eaten off by the rabbits until the roots eventually die (see p. 21). Inside this cage, however, the young *Calluna* shoots produced by the still living roots are protected from rabbit-attack, and they have already grown to a height of 20 cms. and borne many flowers. (See Pl. XII, Photo. 24. Note the great contrast between the vigorous and flowering young *Calluna* shoots inside the cage and the closely nibbled down shoots of *Calluna* outside on the right of the photograph.)

Thus it will be seen that the area of degenerate grass-heath enclosed inside this cage is quickly reverting to *Calluna* heath now that the heavy rabbit "pressure" on the enclosed area is removed.

Agrostis vulgaris and *Rumex acetosella* are also growing much better inside this cage than outside and are flowering fairly profusely, whilst outside no inflorescences of these plants can be found, all of them being eaten down by the rabbits.   Possibly however one of the most interesting things about this cage is that each summer since it was erected it has contained a number of *Campanula rotundifolia* flowers (see Photos 23 and 24) although no other harebell flowers occur on the grass-heath outside this cage for a great distance all around, in spite of the fact that the plant itself is present.   This is a striking additional instance of the especially injurious effects of rabbits upon inflorescences.   These instances make it appear highly probable that the presence of rabbits severely checks the reproduction of many plants by seeds. This particular detrimental influence may well appreciably affect the flora of many regions in England and other countries where rabbits are numerous and ubiquitous.

PLATE XII

Photo. 23. Rabbit-proof cage on middle of the degenerate grass-heath association. Note *Campanula rotundifolia* flowers inside the cage (show whitish) although there are none outside for a very great distance all around. (See page 28.) Note the very vigorous and flowering young *Calluna* stems inside the cage. The grass-heath association inside the cage, which was formerly *Calluna* heath, is very quickly going back to *Calluna* heath now that the rabbit-pressure is removed. (See p. 28 of the text.)

Photo. 24. *Campanula rotundifolia* flowers and vigorous flowering young *Calluna* stems 20 cms. high inside rabbit-proof cage. The large white objects inside the cage are *Campanula* flowers and the small whitish specks are *Calluna* flowers. The large white objects outside the cage are bared flints, many of them being prehistoric flint implements. Outside the cage any *Campanula* inflorescences and *Calluna* stems are nibbled down very closely to the surface of the sandy soil and are quite unable to produce flowers as can be well seen in the photo. (See p. 28.) Note the closely nibbled down and almost flat *Calluna* hummock on the extreme right of the photo. in line with the front of the cage.

EFFECTS OF RABBITS ON THE *CAREX ARENARIA* ASSOCIATIONS.

In addition to attacking the *Calluna* heaths and grass-heath areas rabbits frequently attack the dense *Carex arenaria* associations which are so common in parts of Breckland. Rabbits also attack *Carex arenaria* at Blakeney[1].

When the *Carex arenaria* associations in Breckland are competing with grass-heath (degenerated *Calluna* heath) the *Carex* associations are usually badly attacked by rabbits on their external edges, but when the *Carex* associations are competing with *Calluna* the external edges of the *Carex* associations are not usually much attacked by the rabbits. The probable chief cause of this difference in the rabbit-attack on the external edges, as opposed to internal edges, of the *Carex* associations under these two different conditions are dealt with in detail on pp. 34–40.

Areas bare of *Carex* and often occupied by the typical degenerate grass-heath association and sometimes with decaying fragments of *Calluna* frequently occur around collections of rabbit burrows in the large dense *Carex arenaria* associations (see Photo. 6, Pl. III). The *Carex* shoots around these bare areas are always badly eaten by the rabbits.

Many of these large areas of *Carex arenaria* in Breckland were probably fairly recently typical *Calluna* heaths which have completely degenerated owing to rabbit-attack, and have now become replaced by pure *Carex arenaria* associations, except just around the rabbit burrows where the rabbits, after having caused the *Calluna* heath to degenerate, now keep back from their burrows the otherwise spreading *Carex* (cf. the absence of *Carex* on the rabbit-attacked side of the fence seen in Pl. X, Photo. 20). It was the frequent presence of the typical degenerate heath flora along with decaying remains of *Calluna* around rabbit burrows where the dense *Carex arenaria* is now kept back by the rabbits, that first suggested the probability that many of the *Carex* areas were fairly recently typical *Calluna* heaths.

It has already been stated that *Carex arenaria* very likely became established in this now inland district in ancient post-glacial times when the original bay of the Wash extended near to the western border of the district. The *Carex* has thus probably been in the district from ancient times, and could not have continually spread at its present rate or it would be all over the district by now. The probable explanation of the recent rapid spread of *Carex arenaria* in Breckland is that it is due to the presence of rabbits in the district. Rabbits have only been introduced into England since the Neolithic epoch, and in many instances they have only been preserved from their natural enemies during the last few decades or more recently still. Though they attack *Carex arenaria* they apparently only eat it as a matter of necessity, while they eat *Calluna vulgaris* much more readily. Thus it

[1] **Rowan, W.** "Note on the Food Plants of Rabbits on Blakeney Point, Norfolk." *Journal of Ecology*, **1**, No. 4, p. 274.

comes about that, in a zone where *Carex* and *Calluna* are competing, rabbits confer a great relative advantage upon the *Carex* and enable it to spread, owing to the fact that, although they eat it severely, they eat its competitor to a greater extent. This differential action of rabbits upon competing plants has already been mentioned in connection with the degeneration of *Calluna* heath to rabbit-attacked grass-heath (see p. 24). It is a striking fact that rabbits should be able to confer such very great relative advantages upon certain species within a given sphere of competition even although they may eat down these species very severely.

EFFECTS OF RABBITS UPON THE *PTERIS AQUILINA* ASSOCIATIONS.

The rabbits often also attack some of the fronds of the dense *Pteris* associations which are common in Breckland. They do not appear to like the bracken fronds, which are apparently only attacked when the rabbits are very short of food, or when the alternative food supply consists of some very unattractive plant such as *Erica tetralix*. Only the fronds on the extreme edges of the bracken associations are attacked. The rabbits usually only nibble the softer parenchymatous parts of the frond stalk so that the upper part of the frond topples over, and this usually remains connected with the lower portion of the stalk until the exposed sclerenchyma decays sufficiently for the wind to blow the collapsed upper portion away. Apparently the rabbits only taste the frond stalks, since they usually do not eat them at all extensively even when hard pressed for food. It is very apparent that owing to some cause or other—possibly bitterness or toughness—bracken is not a suitable food for rabbits.

Sometimes, however, when a great depth and area of degenerate grass-heath, associated with many rabbit burrows and innumerable rabbits anxious for extra food, borders upon a bracken association, the total temptation of the rabbits to test the bracken stalks for food is sufficiently great to cause considerable depths of bracken on the edge of the association to degenerate. This has occurred on Cavenham Heath near the spot marked 2 on the map, where the rabbit-attack is extremely heavy.

Usually however the rabbits do not attack the bracken at all extensively, and as the edge of a bracken association advances by rhizome growth, the rabbits which have their burrows on the ground which has recently become occupied by the spreading bracken and which formerly lived on the more attractive vegetation previously occupying the ground are driven to go for their food to the more attractive vegetation on the zone immediately outside the advancing edge of the bracken. The result of this is that if, for instance, the bracken is advancing over *Calluna*, there is usually a bare zone, or a zone of *Calluna* degenerating rapidly owing to attack by rabbits coming from the ground recently occupied by the *Pteris*, just in front of the advancing edge of the *Pteris* association. Thus, especially in the cases of large *Pteris* associations which have already covered a large rabbit-occupied area, the advancing edge of the *Pteris* does not have to compete at all with the *Calluna* which formerly occupied the ground, and for this reason the bracken can advance much more rapidly than if it had had to compete with the original heather. The driven out rabbits eat the heather and thus tend to prepare the way for the advancing bracken. Thus the rabbits through attacking its competitors and scarcely attacking the *Pteris* at all confer a great advantage upon the bracken, which is thereby enabled to spread very rapidly—far more rapidly than it could otherwise do.

When bracken is competing with *Carex*, the presence of rabbits also confers a relative advantage upon the bracken by inflicting a differential rate of injury upon its competitor (as already described) owing to the fact that the rabbits attack the *Carex* more than the bracken; but in this case the advantage conferred on the bracken by rabbits is not nearly so great as where bracken is competing with *Calluna*, since the attack on the *Carex* relatively to bracken is not nearly so great as the attack on the heather relatively to bracken. It is hoped to measure accurately (by means of quadrats and orientated photographs) the relative rates of advance of bracken over *Calluna* and over *Carex* in the presence of rabbits and in the absence of rabbits respectively.

EFFECTS OF RABBITS UPON *JUNCUS* SPP., *SALIX REPENS* AND *ERICA TETRALIX*.

The rabbits also attack various species of *Juncus* on Cavenham Heath (chiefly *Juncus conglomeratus*, *J. obtusiflorus* and *J. supinus*). In some places *Juncus* is attacked very heavily. Often when the *Juncus* borders upon open water the rabbits can only reach the plants on the drier edge of the *Juncus* zone and the plants within reach of the rabbits are usually badly eaten down, all their inflorescences being eaten off, so that they are entirely prevented from flowering, while the plants growing where the water is deeper and thus out of reach of the rabbits produce inflorescences freely.

Sometimes the rabbits also attack young stems of *Salix repens*. They usually only do this however where their alternative food supply is very unattractive.

Erica tetralix is another plant which is not usually much attacked by rabbits at Cavenham. Apparently the rabbits do not like it—possibly this may be largely due to the hairy nature of the plant. *Erica tetralix* is, however, eaten less reluctantly than bracken. The *Erica*, like *Carex*, *Pteris* and *Salix repens*, is eaten chiefly in places where there is no alternative source of food supply, or where the alternative sources of food supply are about as unattractive as these plants are themselves.

Erica, *Carex*, *Pteris* and *Salix*, all rabbit-attacked under these latter conditions, can be well seen in Photo. 2, Plate X. In this instance, the rabbit-attacked *Erica* plants are surrounded, on one side by unattractive *Pteris* and on the other side by unattractive *Salix repens*. Under these special conditions, where the rabbits only have a choice of evils for their food supply, all these very unattractive plants are appreciably rabbit-attacked as can be seen in the photograph, and the *Erica tetralix* bushes have even taken on the particular rounded shape without projecting shoots characteristic of rabbit-attacked bushes. Since the rabbits do not eat *Erica tetralix* nearly so readily as they attack *Calluna vulgaris*, they confer a great advantage upon the *Erica* and enormously increase its possible distribution in a zone where it is competing with *Calluna vulgaris*, in spite of the fact that they do eat it.

This increase in the distribution of *Erica* relatively to *Calluna* and the resulting raising of the *Erica* zone in a *Calluna* heath owing to the differential action of rabbits upon these two competitors has probably a bearing upon the occasional occurrence of *Erica tetralix* in certain very slight depressions on Cavenham Heath only slightly less dry than the surrounding *Calluna* heath itself, as well as in the much damper hollows. For if the distribution of the *Erica* relatively to the *Calluna* has been greatly favoured by the differential action of the rabbits and the *Erica* zone has thus been raised, the successful competition of the *Erica* with the *Calluna* in those places which are only slightly less dry than main *Calluna* heaths themselves may possibly be partly explained.

EFFECTS OF RABBITS UPON *SOLANUM NIGRUM*, *CONIUM MACULATUM*, *URTICA DIOICA* AND *U. URENS*.

Solanum nigrum, *Conium maculatum*, *Urtica dioica* and *U. urens* are the only herbaceous plants which have been found on Cavenham Heath which are not attacked by rabbits more or less severely in one place or another. It has already been indicated that the heavy rabbit-attack by destroying their competitors and practically not eating these poisonous and stinging plants at all must confer a very great advantage indeed upon them. It is very striking to see for instance a herbaceous *Solanum nigrum* plant green and growing vigorously although it is surrounded on all sides by brown, dead and decaying rabbit-attacked *Calluna* and loose *Cladonia*.

EFFECTS OF RABBITS ON SEEDLING TREES RESULTING IN CONFINING TREE GROWTH TO THE VALLEYS.

It has already been stated that various artificial plantations of birch and introduced pines have been made upon the upper portions of many of the Breckland heaths—upon Cavenham Heath for instance—although previously there were no tree plantations in the district. Seeds from some of these plantations have apparently been able to colonise some of the minor valleys.

PLATE XIII

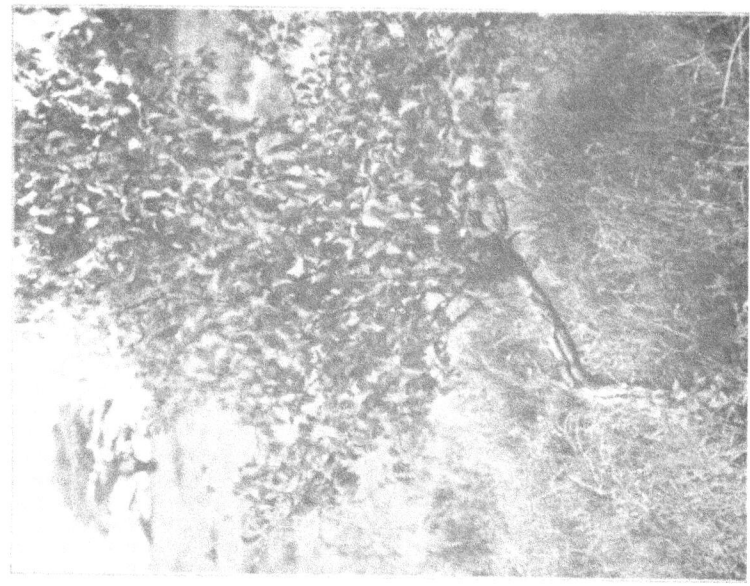

Photo 26. Young *Betula alba* tree on valley side with stem badly attacked by rabbits. Note that the stem has become badly contorted as a result of the rabbit-attack probably owing to uneven release of internal stresses consequent upon uneven eating by the rabbits.

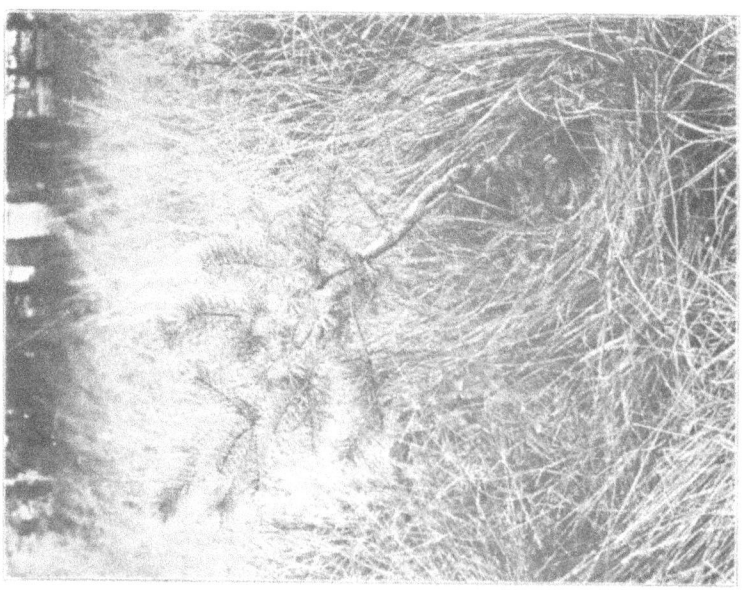

Photo 25. Young *Pinus sylvestris* tree in valley with stem badly attacked by rabbits. The rabbit-attacked stem can be well seen. It is doubtful if this particular tree will be able to survive the rabbit-attack but it may possibly do so. (See p. 33.)

In many of these cases (see Photo. 1, Plate X) there were very few trees in the valleys 50 years ago, but these have apparently produced much seed and they are colonising the surrounding region with many young trees.

In practically all cases the bases of the stems of these young trees are very badly attacked by rabbits (see Pl. XIII, Photos 25 and 26) and in most instances the young trees die from this rabbit-attack. Though practically all the young trees in the valleys are badly attacked by rabbits, and though most of them succumb to the attack, yet some of them do manage to struggle on and survive it, and when they reach a certain height—apparently about 3 feet in the case of *Pinus sylvestris* on Cavenham Heath—their stems become relatively immune to the more serious effects of the rabbit-attack and the trees can then grow with comparative ease.

By far the greater proportion of those young trees which do manage to survive the rabbit-attack to which all of them are subjected, are those situated near the bottoms of the valleys; and as one passes up the sides of the valleys, a greater and greater proportion of the young trees succumb to rabbit-attack, until near the upper edges of the valleys all the young trees eventually succumb before they can reach a size when they would become immune. The rabbit-pressure, as indicated by the number of rabbit dung pellets per unit area, is approximately constant all over the sides of the valleys, and the survival of the young trees on the lower portions of the valley sides is not due to a diminished intensity of rabbit-attack in these zones.

It appears probable that the true explanation is that near the bottoms of the valleys, where there is a much greater available water supply, the young trees can just produce fresh tissue faster than the rabbits eat it and can often just repair the damage caused by the rabbit-attack and survive it. Thus they just manage to grow and struggle on until they reach a size when they become relatively immune and can grow comparatively easily. On the sides of the valleys where the water supply is much less, the growth of the young trees is retarded and they cannot produce fresh tissue as quickly as the rabbits eat it. Thus they are killed by the rabbit-attack before they can reach a sufficient size to become immune.

A portion of the upper area of Cavenham Heath (marked 16 on the map) near the Icklingham road has been fenced off from rabbit-attack and little pine trees are growing inside this rabbit-proof enclosure on the dry upper portion of the heath. In this locality the young pine trees can only grow slowly even when protected from rabbits, yet nevertheless they can grow if protected even on this dry upper portion of the heath, while outside the rabbit-proof enclosure on the dry upper areas, pine seedlings are always killed by the rabbits before they can attain any size.

It is thus seen that the rabbits prevent the dry upper areas of Breckland from being colonised by young pines and from being ultimately converted into fairly useful pinewoods. Such pinewoods would probably be of far more value to this island country than are the rabbits. If it had not been for the largely incidental suppression of the natural enemies of the rabbits consequent upon pheasant preserving, a considerably larger area of Breckland would probably now have been converted into subspontaneous pinewoods.

The limitation of tree growth to the damper valley bottoms and sides consequent upon biotic attack on the seedlings, whereas if it were not for this the tree growth would spread all over the drier upper areas, is an extremely interesting biological phenomenon which probably has a great bearing upon the zonation and distribution of natural vegetation in many parts of the world—for instance in parts of Australia. In this connection it is interesting to note that tree growth on the prehistoric steppes of Continental Europe was apparently largely confined to the valley sides[1]. Many herbivorous animals were present on these prehistoric steppes, and it seems very possible that the limitation of the tree growth to the valley sides was largely due to the action of biotic attack upon the seedlings, as happens in Breckland at the present day. It is clear that the factor of biotic attack must be taken into account, in addition to the water factor, in considering such cases.

ZONATION OF VEGETATION AROUND RABBIT BURROWS IN BRECKLAND.

It has already been mentioned that areas bare of *Carex* often occur around collections of rabbit burrows in the dense and extensive *Carex arenaria* associations in Breckland (see page 29 and also Pl. III, Photo. 6). Some instances were observed, however, on Tuddenham Heath, in which *Carex arenaria* is *associated* with collections of rabbit burrows in *Calluna* heath whilst it is absent elsewhere (see Pl. XIV, Photo. 27). These cases are very interesting because this distribution of *Carex* in relation to the collections of rabbit burrows is quite opposite to the apparently more usual cases in which the *Carex* is *absent* around the collections of burrows whilst it is present elsewhere. In order to obtain information with a view to determining the causes of this apparently reversed distribution of the *Carex* in the two classes of cases the transition zone between the *Carex* association, where it is present immediately around the burrows, and the surrounding *Calluna* heath was first examined (see Pl. XIV, Photo. 28). In passing from the *Carex* zone to the Callunetum the *Carex arenaria* shoots and leaves gradually become shorter, weaker and fewer in number, and occur chiefly along the numerous rabbit tracks (see Pl. XIV, Photo. 28). On the other hand, the *Calluna* bushes gradually become more numerous and of greater height until the typical *Calluna* heath association itself is reached as seen on Photo. 28. The *Calluna* bushes of the transition zone tend to have the rounded form, with no projecting shoots, characteristic of rabbit-attacked bushes. The transition zone is, in fact, a narrow but otherwise typical zone of *Calluna* heath degenerating through rabbit-attack, with the addition that a generating zone of *Carex arenaria* is associated with it, and

---

[1] **Geikie, J.** "The Tundras and Steppes of Prehistoric Europe." *Scottish Geographical Magazine*, 1898.

PLATE XIV

Photo. 27. Area occupied by a relatively small dense *Carex arenaria* association around a collection of rabbit burrows in a *Calluna* Heath.—Tuddenham Heath. The photo. shows a phenomenon which is of very great interest in view of the fact that the *presence* of the *Carex arenaria* around the collection of rabbit burrows and its absence elsewhere appears to be exactly the opposite to the frequent and apparently more natural instances in which the *Carex* is *absent* around the collections of rabbit burrows while it is present elsewhere. (See page 34.)

Photo. 28. Transition zone on edge of a *Carex arenaria* association around a collection of rabbit burrows in a *Calluna* Heath. (Note also *Carex arenaria* around another collection of rabbit burrows in the middle distance.) Note that the *Carex arenaria* shoots and leaves gradually become fewer in number, less in height and weaker in this transition zone, occurring chiefly along the rabbit tracks, while the *Calluna* bushes gradually become greater in height as the typical *Calluna* heath is entered. This is a rather narrow zone of *Calluna* heath degenerating through rabbit-attack but a generating zone of *Carex arenaria* is associated with it. (See p. 34.)

with the exception that, in these special cases, a luxuriant growth of *Cladonia* has not up to the present been found associated with the degenerating *Calluna*. Possibly this may be largely owing to the presence of the generating zones of *Carex arenaria*. A comparison of these degenerating *Calluna* zones with those outside large spreading *Carex* associations on the heaths where the *Calluna* is degenerating and the *Carex* is spreading confirmed the belief that these degenerating *Calluna* zones had similar origins, the degeneration of the *Calluna* and spread of the *Carex* being in both cases due to rabbit-attack on the *Calluna*.

The presence of *Carex arenaria* around some collections of rabbit burrows in the *Calluna* heath, although it was absent elsewhere, and its absence around other collections of rabbit burrows whilst it was present elsewhere, thus appeared to be due to the following series of events. Some rabbits originally made a collection of burrows in the *Calluna* heath and brought about the degeneration of the immediately surrounding *Calluna* (Pl. VII, Photo. 13). After the surrounding *Calluna* had degenerated, *Carex arenaria* colonised this area relatively bare of *Calluna*, the rabbits eating the more attractive surrounding *Calluna* at the edge of and outside the *Carex* rather than the less attractive *Carex* which was colonising the degenerated area itself; until eventually, owing to the differential action of the rabbits on the *Calluna* and the *Carex*, a fairly large area occupied by *Carex arenaria* on degenerate *Calluna* heath around rabbit burrows in the *Calluna* heath was produced (see Pl. XIV, Photo. 27). After the *Carex* had colonised these areas the rabbits continued to go to the edge of the more attractive *Calluna* outside the *Carex* for their food rather than eat the relatively unattractive *Carex* which had colonised the area around their burrows. A degenerating *Calluna* zone outside the edge of a spreading *Carex arenaria* association was thus produced[1] (see Photo. 28) and eventually a fairly large area of *Carex* around the collection of rabbit burrows in the *Calluna* heath resulted (see Photo. 27).

It appeared probable also that, as the degeneration of the *Calluna* through rabbit-attack proceeded, and as the resulting *Carex* association increased in diameter, the rabbits from the central burrows eventually had to go so far to reach the more attractive *Calluna* outside the edge of the spreading *Carex* association that, rather than do this, they tended to eat the relatively unattractive *Carex* because it was much nearer to their burrows, thus eventually producing secondary areas around rabbit burrows bare of *Carex* inside the now larger *Carex* associations.

It appeared that if this theory of the origin of the reversed distributions of the *Carex* in relation to the collections of rabbit burrows is true, smaller areas bare of *Carex* would very likely occur around some of the individual burrows

---

[1] Cf. the production by rabbit-attack of a degenerating *Calluna* zone outside the edge of a spreading unattractive *Pteris* association (see p. 31).

in the collection of rabbit burrows which was, as a whole, associated with *Carex arenaria* in the *Calluna* heath. A search for such smaller incipient bare areas around individual burrows in the smaller *Carex* associations was then made and their actual existence was discovered (see Pl. XV, Photo. 29) although they had not been seen before. The actual existence of these small commencing bare areas around individual burrows affords striking confirmation of the theory of the origin of the reversed and apparently anomalous distribution of *Carex arenaria* in relation to the collections of rabbit burrows.

It may be noted that the close association of the *Carex* (seen on Pl. XIV, Photo. 27) with the collection of rabbit burrows strongly confirms the theory already given (see page 29) that the recent rapid spread of *Carex arenaria* in Breckland and the resulting large *Carex arenaria* associations are associated with and due to the presence of rabbits in the district and their differential action upon the *Calluna* heath and the *Carex*; and do not depend, for instance, upon localised edaphic conditions.

Ordinarily the presence and competition of the *Calluna* prevents the *Carex* from gaining ground on the *Calluna* heaths, but when the *Calluna* heath has partly degenerated from rabbit-attack, *Carex arenaria* can colonise the degenerated area if its seeds are present, the rabbits eating the *Calluna* on the degenerating edge outside the *Carex* association rather than the relatively unattractive *Carex* in the centre. When the *Calluna* has degenerated to a considerable distance from the burrows, the rabbits, rather than go this distance to the *Calluna*, tend to eat the relatively unattractive *Carex* immediately around their burrows, thus producing secondary areas bare of *Carex* around the burrows although the *Carex* is associated with the burrows as a whole.

Thus the reversed and apparently anomalous distribution of the *Carex* in relation to the rabbit burrows is explained. It is the relative distance of the more attractive *Calluna* with resulting variation in the rabbit-pressure on the *Carex* which is the determining factor in the two classes of cases. Both conditions are really different stages fairly wide apart in the same biotic succession. These phenomena are illustrated in the accompanying diagram (Fig. 6). The apparently irregular case of the presence of *Carex* immediately around the collection of burrows and its absence elsewhere (see Photo. 27) is developmentally a more primitive condition than the more obvious and apparently more natural case of its absence around the burrows and its presence elsewhere; but when, as often happens, the more attractive *Calluna* has degenerated to a considerable distance before *Carex* has colonised a particular collection of burrows, the primitive condition of close association of the *Carex* with the collection of burrows may never occur, owing to the rabbit-pressure on the available food being sufficiently great to keep the spreading *Carex* back from the collection of burrows from the first.

When the rabbit-pressure is very great, as on Cavenham Heath, and the

Calluna heath is degenerating rapidly, the rabbit-pressure is often also sufficient to keep down the *Carex* shoots and practically obliterate the normally intermediate *Carex arenaria* zone which would otherwise become apparent, so that the developmentally intermediate *Carex arenaria* zone is practically

Fig. 6. The Biotic Zonation of Breckland—Zonation of Vegetation around Isolated Rabbit Burrow. For further description see text, pages 34—40.

absent and the *Calluna* heath appears to degenerate directly to grass-heath. The same effect is observed under conditions of less heavy rabbit-attack when *Carex arenaria* seeds and rhizomes are absent, i.e. in those regions which the spreading *Carex arenaria* has not yet reached.

THE BIOTIC ZONATION OF BRECKLAND.

Examination of the accompanying photographs (Photos 27, 28 and 29) and the explanation which is given above will make it clear that the effect of the rabbits is to produce a particular zonation in the vegetation immediately around their burrows as is illustrated in the accompanying diagram (Fig. 6, p. 37). The centrifugal order of the various vegetation zones in this zoned vegetation around rabbit burrows in Breckland is as follows: 1. (bare sand and lichens), 2. grass-heath zone, 3. *Carex arenaria* zone, 4. typical heath zone (*Calluna* etc.), 5. tree zone (pine woodland, etc.).

This typical zonation of vegetation around rabbit burrows on otherwise uniform and level sandy tracts in Breckland is a very beautiful and striking example of a dynamic vegetation succession or developmental series of conditions produced by the differential effects of the diminishing intensity of biotic attack upon various types of vegetation as the distance from the rabbit burrows increases.

If the intensity of the radiating rabbit-attack from the burrows gradually increases the various zones in the zoned vegetation around them would gradually expand outwards (as is indicated by the arrow heads in Fig. 6) until eventually practically the whole of the area associated with the burrows would become occupied by dwarf grass-heath vegetation.

If, on the other hand, the radiating intensity of the rabbit-attack from the burrows gradually diminished the various zones in the zoned vegetation around them would gradually contract and close up to the centres until eventually practically the whole of the associated sandy area would become occupied by pine woodland.

It is thus seen that variation in the intensity of the rabbit-attack alone is sufficient to change the dominant type of vegetation in Breckland from pine woodland to dwarf grass-heath vegetation through the phases of *Calluna* heath and *Carex arenaria*, and that for each given intensity of rabbit-attack there is a certain associated vegetation. In other words, if the rabbit-attack were non-existent or very slight, pinewood would be the dominant type of vegetation on a particular area. If the intensity of the rabbit-attack were slightly greater and remained at a certain constant level *Carex arenaria* would tend to be the dominant type of vegetation on the area, and if the intensity of the rabbit-attack were much greater still, grass-heath would tend to be the dominant type of vegetation on the area.

It should be noted however that the presence of *Pteris aquilina* in any locality interferes with the typical biotic zonation of Breckland, for although *Pteris* cannot dominate grass-heath under conditions of extremely heavy rabbit-attack yet it can dominate *Carex* and *Calluna* under all intensities of rabbit-attack.

The following theory to account for this production of zoned vegetation around rabbit burrows in Breckland may possibly be interesting. Suppose the amount of rabbit-attack emanating from the burrows to be gradually reduced to zero, then the *Carex arenaria* of the *Carex* zone which can grow taller than the grasses would gradually invade and smother the more dwarf grasses of the grass-heath zone, and the central grass-heath zone would eventually disappear. The *Calluna* of the *Calluna* zone which can grow taller than the *Carex* would gradually invade and smother the more dwarf central *Carex* zone and the central *Carex* zone would eventually be obliterated. In the same way the pine trees of the outer pinewood zone would gradually colonise the central *Calluna* zone and would eventually smother the more dwarf *Calluna*[1]. Probably the enormous advantages that plants which can grow tall possess in being able to dominate and often eventually to exterminate more dwarf competitors largely no doubt owing to interception of the light are not sufficiently emphasised in the existing literature relating to vegetation. Although tall growing plants possess enormous potential advantages over more dwarf competitors yet it will be noticed that all the different kinds of plants concerned in this zonation accord with the generalisation already given on page 27, that taller growing plants eventually suffer more from increasing biotic attack than do more dwarf growing competitors. Increasing biotic attack has the effect of an increasing downward pressure on associated vegetation. It will be noticed that there is an upward gradation in height of the different types of vegetation as the distance from the rabbit burrow increases. Suppose now the biotic attack from the burrows to rise gradually from zero to a maximum. The pinewood round the burrows would eventually disappear owing to death of the old trees and continual destruction of the individual seedlings. The resulting open and well-lighted space could eventually become colonised by the more dwarf *Calluna* which does not suffer so much from the biotic attack as do the seedling trees. When the rabbit-attack immediately around the burrows became greater the *Calluna* heath itself in this position would degenerate and would be replaced by the more dwarf *Carex arenaria*. As the biotic attack from the burrows gradually increased the above zones would gradually expand outwards, and the central portion of the *Carex arenaria* association immediately around the burrows would eventually degenerate to still more dwarf grass-heath. Thus the characteristic biotic succession and zonation of the vegetation around rabbit burrows in Breckland would be produced. It will be seen that, if the above theory is true, this interesting zonation largely arises owing to the opposition between the more injurious effects of biotic attack upon taller growing plants and the advantages which taller growing plants would normally possess over more dwarf growing competitors, and that the conditions of the zonation at any time largely represent the effects of a balance between these two opposing

---

[1] Cf. the occurrence of *Calluna* in well-lighted spots in woods. **Tansley, A. G.** *Types of British Vegetation*, 1913, p. 99.

Influences. Probably there are many other zonations and successions in natural vegetation which depend for their existence upon the opposition and balance between the more injurious effects of biotic attack upon taller plants and the otherwise natural advantages of tall growth. . *Indeed the zonation and present dominance and distribution of the different types of natural and semi-natural vegetation in the various parts of the uncultivated portions of the British Isles and of other countries which have been subjected to a large amount of biotic attack by grazing animals may largely depend upon variations in the intensity of the biotic attack in different regions producing variations in the effects of the balance between the progressively more injurious effects of increasing biotic attack upon the taller kinds of plants and the great natural advantages which the taller growing plants would otherwise possess over their more dwarf competitors.*

Summary and General Remarks on the Effects of Rabbits upon Vegetation.

It is thus seen that in addition to causing the main *Calluna* heaths to degenerate to grass-heaths, which rapidly revert to Callunetum on protection from rabbits, these animals also produce various other striking effects upon the vegetation. One of their most important effects is that they are especially injurious to taller growing plants and to inflorescences. *Campanula rotundifolia* and many other plants are entirely prevented from flowering.

The rabbit-attack tends to reduce greatly the number of species of plants present on any area and is especially destructive to various dicotyledonous species, thus tending to favour the growth of grasses relatively to dicotyledons.

The rabbit-attack also limits tree growth to the damper valleys where the young trees can grow comparatively vigorously while if it were not for the rabbit-attack tree growth could spread over and dominate all the upper areas.

In addition to attacking the *Calluna* heaths, grass-heaths, and young trees the rabbits also attack the large *Carex arenaria* and *Pteris aquilina* associations which are common in Breckland, and the possible and actual distributions of all these plant associations are very greatly modified by the differential effect of the various intensities of rabbit-attack upon the various competing species. Indeed the chief characteristic of the vegetation of Cavenham Heath is the extreme mobility of the various plant associations due to the differential action of varying intensities of rabbit-attack upon them in upsetting the balance and thus altering the ultimately dominant types of vegetation.

The vegetation of Blakeney Point is also considerably more mobile than might have been suspected[1]. Probably the changes in the vegetation due to topographical changes are more rapid at Blakeney Point than on Cavenham Heath owing to the presence of mobile shingle banks and sand dunes in the former case; but on the other hand, the general changes in the vegetation are

---

[1] **Oliver, F. W.** "Blakeney Point in 1913." *Journal of Ecology*, **1**, 1913, p. 4.

far more rapid on Cavenham Heath than on Blakeney Point owing to the greater operation of biotic factors on Cavenham Heath which completely outweighs the comparative absence of topographical changes.

The different intensities of the biotic attack at varying distances from the rabbit burrows produce a characteristic zonation of the vegetation around the burrows on the otherwise uniform sandy plains of Breckland. This zonation is a very striking and beautiful example of a dynamic biotic succession depending for its existence upon the different amount of biotic attack at various points. The existing dominances of the different kinds of vegetation on their various zones around the burrows, and also the existing dominances of the corresponding kinds of vegetation on the ground occupied by the corresponding large *Calluna* heaths, *Carex arenaria* and grass-heath associations of Breckland, ultimately depend for their existence and maintenance upon the existence and maintenance of different intensities of biotic attack upon the vegetation of the corresponding areas, and the respective dominances of these different kinds of vegetation do not depend upon specially localised differences in the soils.

The very complex distributions of these various plant associations on different areas in Breckland was for long unexplained, but we now know that they primarily represent individual developmental phases in a particular degenerative succession of taller vegetation due to the varying intensity of biotic attack at different points[1].

Ecological factors are sometimes grouped solely as climatic and edaphic, but the highly important biotic factors which are capable of exerting such great influences upon vegetation should probably always be included in the classification, and the extent to which they are or have been present in particular cases and their effects should probably always be considered when dealing with the existing zonation and distribution of natural and semi-natural vegetation in any district. Apparently the presence of rabbits alone is sufficient to change the dominant vegetation of Breckland where *Pteris* is absent from pinewood through *Calluna* heath and *Carex arenaria* associations to dwarf grass-heath.

Note on the Effects of Rabbits upon the Vegetation of West Newton Heath, near Sandringham, in North-west Norfolk.

Although rabbits often produce areas bare of *Carex* around collections of their burrows in the dense *Carex arenaria* associations of Breckland, the writer has not yet seen areas bare of the highly unattractive *Pteris* around individual collections of rabbit burrows in the *Pteris* associations of Breckland. Apparently the rabbits eventually prefer to migrate right outside of the

---

[1] It is instructive to realise that much time might have been wasted in the first place in examining and analysing the soils under the different plant associations in Breckland on the supposition that the differences in the vegetation were due to differences in the soils. Various

spreading *Pteris* associations to a region with more attractive food, rather than to eat the unattractive *Pteris* when it has spread a long way all around their burrows. In Breckland, they are apparently always eventually able to find a more suitable region with more attractive food to migrate to and so are able to avoid ultimately having to eat the unattractive *Pteris*.

The writer has however seen areas bare of *Pteris* around rabbit burrows in a *Pteris* association on West Newton Heath near Sandringham in North-west Norfolk (see Pl. XV, Photo. 30). The rabbits are much less numerous per unit area on this heath than they are on most of the Breckland heaths, but in spite of this, if they are so hungry that they have to eat the very unattractive *Pteris* around their burrows in this locality, the actual rabbit *pressure* on the vegetation of West Newton Heath must be far greater than that on the Breckland heaths where the phenomenon does not occur. If, in spite of the relatively few rabbits, the actual rabbit-pressure on the vegetation of this North-west Norfolk heath is far greater than that on the Breckland heaths, the results of rabbit-attack observed in Breckland may have a wider application than might otherwise have appeared to be the case. Probably the heavier rabbit-pressure at the present day on the vegetation of West Newton Heath largely arises out of the fact that most of the vegetation now consists of unattractive *Pteris aquilina* and of unattractive *Erica tetralix* so that the rabbits have to eat these plants, and any more attractive plants which remained would probably be very quickly suppressed. The extensive distribution and the great purity of the unattractive *Erica* and *Pteris* associations on West Newton Heath at the present day are thus probably largely due to the cumulative effect of rabbit-pressure in bringing about the degeneration of any more attractive previous plant associations and in ultimately severely eating off more and more rapidly as they became rare any isolated relatively attractive plants amongst the mass of unattractive vegetation.

differences in the water contents, etc., of the different soils would undoubtedly have been detected, but these particular differences would have been largely consequent upon the occupation of the soils by different kinds of vegetation (owing to differing intensities of biotic attack) and not *vice versa*. These considerations emphasise the advisability when beginning an ecological study of a particular area of spending at first a considerable time in looking round and collecting general information before embarking on laborious work on the basis of merely preliminary hypotheses.

PLATE XV

Photo. 29. Small incipient grass-heath area bare of *Carex* around individual rabbit burrows in an area of *Carex* which is, as a whole, associated with rabbit burrows in *Calluna* heath. The discovery of these small incipient bare areas around individual burrows affords striking confirmation of the theory relating to the apparently anomalous and reversed distribution of the *Carex* in relation to the collections of rabbit burrows. (See p. 36.) Compare the zonation of the vegetation seen in the picture with that indicated in the diagram, Fig. 6, p. 37 in the text.

Photo. 30. Area bare of *Pteris* around rabbit burrows in a *Pteris* association on West Newton Heath, near Sandringham. An area associated with rabbit burrows and almost devoid of *Pteris* fronds can be well seen in the photo. This phenomenon is rare in East Anglia and the writer has not yet seen it in Breckland. (See p. 42 of the text.)

CHAPTER IV.

EXPERIMENTS MAINLY RELATING TO THE AVAILABLE WATER SUPPLY.

METHODS OF INVESTIGATION.

When the great influence of rabbits upon the vegetation of the Breckland heaths had been ascertained (see Chapter III) it became clear that unless the rabbit pressure were removed from an area the rabbits would spoil the results of any detailed experiments upon other factors.

A large quadrat, 3 metres square, in the middle of the degenerate grass heath association was therefore carefully fenced off from rabbit attack by wire netting (on April 5th, 1914). The rabbit-proof wire netting of this large quadrat projects 4 feet above the surface of the soil and is inclined slightly outwards (see Pl. XVI, Photo. 31, and Fig. 7, p. 46). The netting was let down several feet into the sandy soil and the bottom edge was bent sharply outwards, so that if any rabbits tried to burrow down immediately outside the netting they would eventually encounter netting on two sides and would probably desist from further effort.

After the discovery of the great effects of the rabbits on the vegetation it was thought that the poorness of the vegetation of the grass heaths was probably chiefly due to the rabbits, and it was expected that when all the rabbits were removed from an area the associated vegetation would quickly become much more luxuriant.

This however proved not to be the case during the first year. The vegetative portions of the plants inside the large rabbit-proof quadrat were but little more luxuriant, and only slightly taller, than those of the plants on the exposed grass heath, the chief difference being that the sheltered area bore a far greater number of inflorescences.

It thus appeared that the growth of the vegetation on this area was probably being severely[1] limited by some other factor or factors than

---

[1] The word "severe" is used in this connexion to mean that the rate of growth is very much less than it would be if the value of the factor were so increased that it ceased to be a limiting factor: or, to put the matter another way, that its value would have to be greatly

rabbit attack. It appeared that the rabbit attack probably tended to control the *luxuriance* of the vegetation (i.e. the amount of vegetative substance present), tending to keep it down to a certain small amount, while it also appeared that the *rate of growth* of the vegetation was probably being severely controlled by some other factor[1].

General observations upon the usually greater luxuriance of the vegetation in the valleys when protected from rabbit attack and where the soil is much damper—although the peaty soil of some of the valleys of course introduces other edaphic conditions—compared with the lack of luxuriance of the vegetation on the upper areas when protected from rabbit attack, and general considerations upon the porous nature of the sandy soil of the upper areas and the low rainfall of the district, made it appear that the available water-supply might be an important limiting factor to the growth of the vegetation on the upper areas of the district.

It was at first intended to estimate the holard and chresard of the soils on the upper areas and of the soils in the valleys by one of the usual laboratory methods. There are however two serious objections to such methods when employed as a means of determining the actual factors at work on natural vegetation. In the first place the sources of error due to the fluctuating water content of different layers and of adjacent masses of soil, to determinations taken at different periods after rainfall, etc., are numerous. And even if these sources of error are recognised and corrected the supposed information as to the ecological factor at work would depend upon the assumption that the observed differences in the vegetation are due to water supply, whereas they might in reality be due to some wholly different factor, such as variation in some mineral constituent of the soil.

It thus appeared that actual experimentation by alteration of the existing

increased before the factor ceased to limit the rate of growth. The extent of the "severity" at any instant of time would be measured by the difference between the rate of growth if the value of the factor in question were so increased that it ceased to be a limiting factor, and the otherwise normal rate of growth; and the total effect of the severity during any given period of time would be measured by the area enclosed between the graphs of these two functions.

[1] The relation between luxuriance, rate of growth and rate at which the rabbits eat the vegetation may be expressed mathematically as follows:

If   Rate of growth of the vegetation $= V$

Rate at which the rabbits eat the vegetation $= R$,

and   Amount of vegetation existing at any time, i.e. luxuriance $= L$.

Then   Rate of increase of L at time $t = \dfrac{dL}{dt} = V - R$.

For the condition of equilibrium to be stable the condition required is that if L is greater than its equilibrium value the sign of $\dfrac{dL}{dt}$ must be negative, and where L_0 is the initial value the value of L after a finite time t will be

$$L_0 + \int_0^t (V - R)\, dt.$$

water supply of the vegetation itself on the upper areas would be a far sounder method of determining whether and to what extent available water supply was an effective limiting factor to the growth of the vegetation on these grass heaths than mere direct determinations of water content.

WATER DRIP AND MANURING EXPERIMENT.

The method it was eventually decided to employ in this instance was to increase artificially the available water supply at a certain spot by means of a water drip from a water-containing vessel; and if the actual growth of the vegetation were increased at this spot it would be established that under the natural conditions the available water supply was an effective limiting factor to the growth of the vegetation on the heath[1]. Further the magnitude of the observed effect might give some indication of the extent of its actual limiting influence under the natural ecological conditions. A barrel fitted with an ordinary wooden tap proved quite a satisfactory water container provided the tap was protected from the heat of the sun.

It was also thought that the available supply of some manurial constituent might possibly be an effective limiting factor to the growth of the vegetation. A metre quadrat was therefore heavily manured with farmyard manure to see if this resulted in an increased growth of the vegetation under the dry conditions. In order to ascertain any combined effects, in addition to the separate effects, of increased water supply and increased supply of manurial constituents, the water drip was arranged so as to add to the water supply at a point half way along the line separating a manured quadrat from an unmanured quadrat (see diagram, Fig. 7). The water was arranged to drip fairly profusely in order that the water supply should cease entirely to be a limiting factor immediately around the drip; and since it was desired that any result should allow for the total time as well as for the absolute extent to which the available water supply might be a limiting factor to the growth of the vegetation and should indicate the product of these two quantities, the water was arranged to drip continuously.

After the water drip had been working for about a week, it was noticed that the grass for a radius of about 20 cms. all around it was slightly taller than the grass elsewhere (see Fig. 7). The increased height was at first about uniform over this circular area around the water drip and fell off very rapidly at the edge of the circle. As time went on however the grass near the centre of the circle close to the actual drip increased in height more quickly than that near the outer edge of the—at first uniform—irrigated circle, and eventually the vegetation immediately around the drip became very much taller and more luxuriant than the vegetation elsewhere (see Pl. XVI, Photos 31 and 32).

[1] Cf. **Blackman, F. F.**, and **Smith, A. M.**, "Experimental Researches on Vegetable Assimilation and Respiration IX," *Proceedings of the Royal Society*, B, vol. **83**, 1911.

As a result of this experiment there is no doubt whatever that the available water supply is a factor which severely limits the growth of the natural vegetation of this grass heath, and the very great importance in ecology of the conception of limiting or controlling factors is strikingly illustrated.

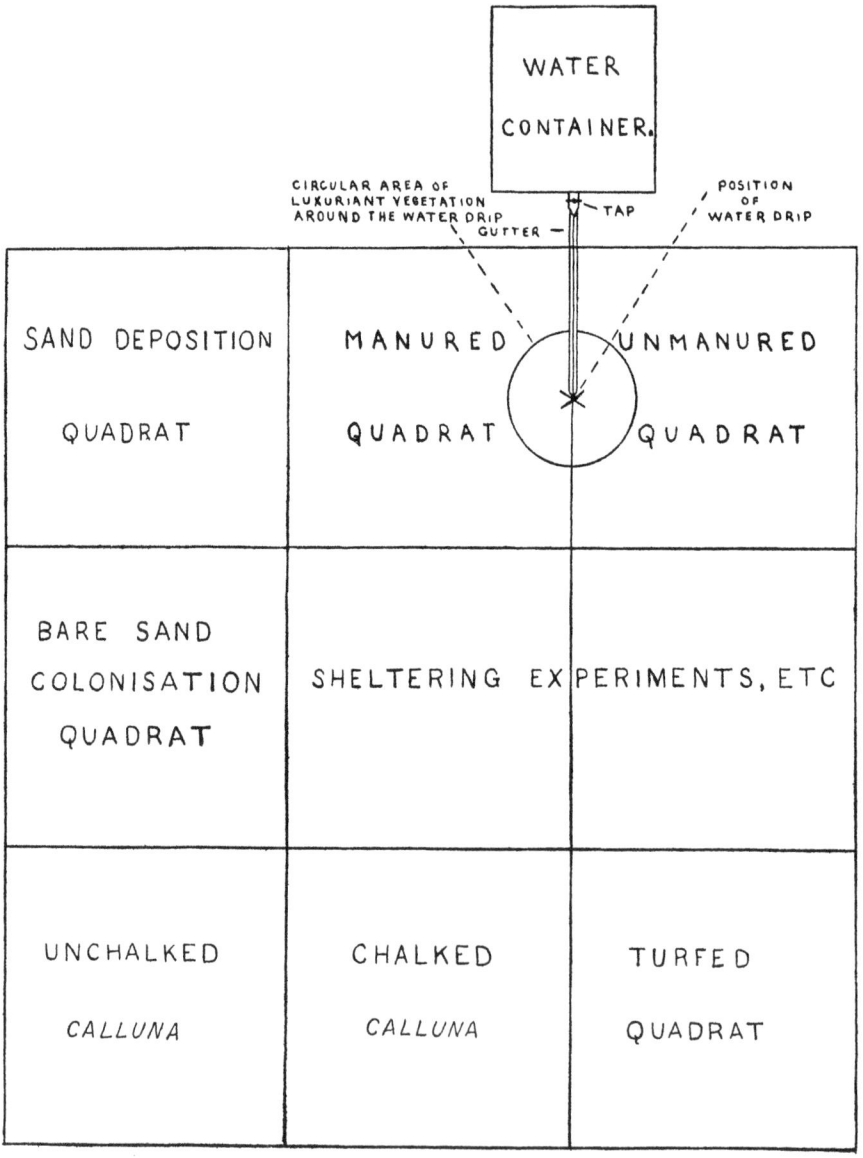

Fig. 7. Diagram showing arrangement of the experimental quadrats, etc. inside the large rabbit-proof enclosure. Each minor quadrat is 1 metre square. See text for description of the water drip, of the manured and unmanured quadrats, and of the sheltering experiments. The other experimental quadrats will be dealt with in subsequent parts.

PLATE XVI

Photo. 31. Large experimental quadrat (3 metres side) fenced from rabbit attack by wire netting and situated on the rabbit eaten grass heath association. Various experiments have been made upon the protected vegetation inside and the results are described in the text. (Cf. Fig. 7, p. 46.) This enclosure was laid out and fenced in April, 1914 and the photograph taken in August of the same year.

Photo. 32. Limiting factor experiment on the available water supply of the protected grass heath vegetation inside the large rabbit-proof enclosure. The continuous water drip from the barrel resulted in greatly increased height and luxuriance of the vegetation immediately around the drip, thus definitely proving that under the natural conditions the deficient water supply is a severe limiting factor to growth (p. 45). The increased luxuriance of the vegetation on the manured quadrat between the drip area and the boarded quadrat on the left can also be seen (p. 47). Note too the increased luxuriance along the base of the wire netting (cf. Photos 31, 33 and 34). A 30 cm. rule is seen on the right of the drip area. The small wooden and glass quadrats and small "greenhouse" referred to in the text (p. 50) can also be seen in the photograph. Photographed August 1914.

The result of this irrigation experiment emphasises the advisability of doing everything possible to conserve the natural water supply if ever these heaths are reclaimed for the purpose of arable cultivation.

It has already been mentioned that the irrigation at first produced a circular patch of vegetation of a uniformly increased height which fell off rapidly at the edges, but that afterwards the more central portions grew taller than the outer portions of the at first uniform patch. The probable explanation of this difference is that at first the whole of the vegetation of the uniform patch received water as quickly as it could be utilised by increased transpiration and growth, but that, as the central plants grew taller and taller, having the first access to the increased water supply, they were gradually able to use a greater and greater proportion of it, so that there was then less left for the outer plants, which consequently could not then grow as fast as the central portions, although they did so at first.

It has also been mentioned that a quadrat on one side of the water drip was heavily manured whilst a quadrat on the other side of the water drip was left unmanured. Soon after the experiment had been started, it was noticed that the vegetation of the manured quadrat was slightly more luxuriant than that of the unmanured quadrat (see Pl. XVI, Photo. 32, and Fig. 7, p. 46). (This also applied to the vegetation of a manured quadrat which was exposed to rabbit attack outside the large rabbit-proof quadrat, but in this case the rabbits quickly nibbled off the slightly more luxuriant vegetation on the manured area, and soon brought it down to the low constant level of the general vegetation of the exposed grass heath.)

With regard to the manured quadrat protected from rabbit attack one was rather surprised at first that the addition of the manure alone produced any increase in the luxuriance, since the water drip experiment was showing that the available water supply itself was such a severe (i.e. powerful and long continued) limiting or controlling factor to the growth of the vegetation. The explanation no doubt is that in nature there are a number of different limiting factors operating at different times. Probably in this special case for instance, after rain the available water supply ceases for a time to be the limiting factor to the growth of the vegetation and something else becomes the limiting factor—probably in this case chiefly the available supply of potassium. It is hoped to test this latter supposition later on by the experimental addition of potassium salts without placing faith in laboratory determinations of so-called "available" potassium. In this connexion the greater height of the tied-up sheaf of grass on the manured side of the water drip as compared with that on the unmanured side should be noted (Pl. XVI, Photo. 32. A 30 cm. rule is seen in the photograph).

Length of Leaf Blades of *Agrostis vulgaris* in cms.

		Maximum	Average	Average Increase
1.	Untreated Area	3	1·5	
2.	Manured only	12	9·0	7·5
3.	Irrigated only	47	36·0	34·5
4.	Manured and irrigated	59	45·0	43·5

From the above figures it will be seen that the water drip alone produced an average increase of 34·5 cms. in the length of the leaf blades of *Agrostis vulgaris* while the addition of manure alone only produced an increase of 7·5 cms. It will thus be seen that it has been experimentally demonstrated that water supply is a far more severe limiting factor to the natural growth of the vegetation on this grass heath than anything which was supplied in the complete general manure.

The vegetation of the irrigated patch consists almost entirely of *Agrostis vulgaris* although on the untouched grass heath *Festuca ovina* is co-dominant with *Agrostis vulgaris*[1]. Apparently the alteration of one factor of the environment alone has enabled the *Agrostis* to crowd out the *Festuca*. As the experimental alteration of the water supply alone has enabled the *Agrostis* to become dominant on this dry sandy heath this may rather tend to indicate that the far greater dominance of *Agrostis* on the damper siliceous soils is very likely really largely due to the greater water supply. This matter should however be directly tested experimentally in some way or other and it may justly be hoped that the employment of such experimental methods in ecology will contribute far more materially to the sound advance of ecology than the laboratory determinations and deductive processes which are frequently solely employed. It will be seen later that the incomplete dominance of *Agrostis* on Cavenham Heath is also partly due to the heavy biotic attack on this area.

"EDGE-EFFECT."

In addition to the luxuriance of the vegetation being greatly increased just around the water drip there is also a strip of vegetation more luxuriant than the general vegetation of the enclosed area just inside the wire netting along the sides of the quadrat (see photographs).

The writer has observed during rain that some of the rain water falling on the sloping wire netting fence tends to be collected by the sloping wire netting, to trickle down it and to be deposited along a line at the bottom of the netting on the surface of the soil: probably the same thing happens with dew. The greater luxuriance of the vegetation just inside the wire netting around the quadrat is thus probably largely due to the netting

---

[1] It will be noticed that the leaf blades of the *Agrostis* on the irrigated patch are badly laid (Photo. 32).

collecting rain and dew and thus adding to the available water supply of the strip of vegetation just where the netting comes in contact with the soil.

The greater luxuriance of the vegetation just inside the edge of the large rabbit-proof quadrat compared with the central enclosed vegetation is however probably chiefly due to what may be termed an "edge effect." The original vegetation just outside the quadrat was unavoidably destroyed in the digging necessary to insert the netting in the soil, and in any case the rabbits would keep any external vegetation nibbled down closely to the surface of the soil. The sparse and short external vegetation cannot transpire so much water per unit area of soil as the more luxuriant internal protected vegetation, and the roots of the strip of protected vegetation just inside the wire netting and bordering upon the sparse external vegetation can thus absorb some of the water supply which would otherwise be used up by the latter. This external strip of protected vegetation can thus grow more luxuriantly than the central area of protected vegetation which has to compete for its water supply with equally luxuriant vegetation on all sides[1].

Wooden boarded quadrats have been let into the surface of the soil along one side of the large rabbit-proof quadrat for the purpose of various detailed experiments upon the vegetation, which will be dealt with later. Along the border of one of these, which had been denuded of vegetation, the "edge effect" mentioned above was very strikingly evident.

EFFECT OF WINTER RAINS ON PROTECTED VEGETATION.

It was mentioned on p. 43 that when all the rabbit pressure was removed from the area inside the rabbit-proof enclosure the associated vegetation was expected rapidly to become considerably taller and more luxuriant than the vegetation outside. This however proved not to be the case to any extent during the first year, except just inside the sloping wire netting and of course around the water drip.

Since the experiment upon the available water supply was showing that water supply was severely limiting the growth of the vegetation it was thought that after the following winter rains the protected vegetation would probably become considerably more luxuriant than during the first year of the experiment. This expectation was realised, for in the second year the general protected vegetation inside the large rabbit-proof enclosure was considerably more luxuriant than it was during the first season (compare Photo. 33 with Photo. 32). *Agrostis vulgaris, Festuca ovina, Aira praecox, Rumex acetosella, Carex arenaria* and *Calluna vulgaris* especially became

---

[1] Cf. **Shantz, H. L.**, "Plant Succession on Abandoned Roads in Eastern Colorado." *Journal of Ecology*, **5**, p. 25. The larger size and greater luxuriance of plants growing on the edge of areas cleared for crops and the bending of the roots of these plants under the open ground, i.e. in the direction of greater water supply, are very strikingly seen everywhere in the short-grass vegetation of the dry Colorado Great Plains.

much more luxuriant and flowered much better inside the rabbit-proof enclosure in the second year than during the first year (see Plate XVII, Photo. 33). Outside the enclosure, on the exposed grass heath, no inflorescences of these plants occur except occasional very dwarf ones of *Aira praecox*. *Galium saxatile* is seen in Photo. 33 flowering luxuriantly in the S.W. corner of the quadrat and also a short distance along on the right-hand side. Outside this quadrat on the general exposed surface of the grass heath no *Galium saxatile* flowers were to be seen. On the exposed grass heath the rabbits entirely prevent the flowering of this plant as well as of the species previously mentioned, viz.: *Campanula rotundifolia, Sedum acre, Rumex acetosella, Galium verum, Calluna vulgaris, Festuca ovina* and *Agrostis vulgaris*. (See p. 27.)

OTHER EXPERIMENTS AND OBSERVATIONS.

Later on in the second year the *Agrostis vulgaris* on the manured quadrat produced a very large number of inflorescences in a dense mass and of an average height of 35 cms. These inflorescences were enormously more numerous and of far greater height on the manured quadrat than elsewhere, with the result that the manured quadrat presented a very striking contrast to the rest of the enclosure. This striking effect of the manure upon the production of inflorescences by *Agrostis* did not occur during the first year.

Shortly after the large rabbit-proof quadrat had been constructed a small wooden quadrat was pegged down on the enclosed grass heath in order to shelter partially an area of the vegetation from sun and wind, and small quadrats with glass sides were also pegged down in order to shelter areas from the wind while leaving them exposed to the sun. The small glass quadrats were made very shallow so that they should not give any results through trapping radiant heat. Small wooden boards and glass sheets were also fixed at right angles on the grass heath in order to shelter areas from wind and sun and from wind only respectively. A small sheet of glass was laid on the surface of the grass heath and a straw mulch was pegged down on another area of the surface. In addition a small glass greenhouse with sloping sides and open ends was constructed with two sheets of glass. Some of these devices can be seen in Photos 31, 32 and 33. None of these arrangements gave any very definite result during the first year except that the vegetation under the piece of glass which was laid on the surface of the grass heath very quickly died, and that the vegetation just behind the northern side of the board which ran from east to west and which was partly sheltered from the sun was appreciably more luxuriant than the general enclosed vegetation. In the spring of the second year, however, the vegetation under the small glass greenhouse consisted entirely of a dense mass of *Galium saxatile* which flowered very vigorously and completely filled the glass shaded

PLATE XVII

Photo. 33. Large rabbit-proof quadrat in spring of 1915 (second year of experiment) showing the greater luxuriance of the enclosed vegetation after the winter rains (p. 49). Note the extra luxuriance of the vegetation along the base of the wire netting (p. 48). *Galium saxatile* is seen in full flower in the S.W. corner. No inflorescences occur on the exposed grass heath outside the enclosure (p. 50). Photographed June 1915.

Photo. 34. Young *Calluna* shoots growing and flowering vigorously inside the large rabbit-proof quadrat in autumn of second year. Some of these have grown over 40 cms. high (p. 50) in a little over a year now that they are protected from rabbits and under the influence of increased water supply available along the base of the netting. Photographed August 1915.

space. In the second year *Poa pratensis* appeared inside the large rabbit proof enclosure near the N.W. corner and flowered vigorously. This plant had not previously been noticed on Cavenham Heath and it is not known how or why it has appeared. Possibly some of the seeds may have been introduced with the wire netting or with the stakes.

During the first year *Festuca ovina* was co-dominant with *Agrostis vulgaris* in the grass heath vegetation inside the large enclosure, but during the second year *Agrostis vulgaris* was far more prevalent than *Festuca ovina*. Hence it appears that *Agrostis vulgaris* suffers more from heavy rabbit attack than does *Festuca ovina*. This is probably largely because the *Agrostis* leaves tend to grow taller than the *Festuca* leaves and are thus more readily eaten and suffer more from the rabbit attack, but when the heavy rabbit pressure is removed the taller growing *Agrostis* leaves are able to smother the more dwarf *Festuca* leaves. (See pp. 27 and 39.)

During the second year as well as during the first year the vegetation just inside the sloping wire netting was considerably more luxuriant than the general enclosed vegetation. The difference in the luxuriance of the protected vegetation near the bottom of the wire netting and the general central internal protected vegetation in the second year can be seen in Plate XVII Photo. 33, in which *Agrostis vulgaris*, *Festuca ovina*, *Aira praecox*, *Rumex acetosella*, *Galium saxatile*, *Luzula campestris*, *Calluna vulgaris*, and *Carex arenaria* are growing and flowering vigorously.

Very vigorous and luxuriant young flowering *Calluna* stems growing on the protected side near the bottom of the wire netting in the second year are especially well seen in Photo. 34. The young *Calluna* stems in this position have grown over 40 cms. high in a little over a year, whilst outside the rabbit-proof enclosure any *Calluna* stems are eaten down by the rabbits almost to the level of the sandy soil and are entirely prevented from flowering.

The luxuriance of the protected vegetation seen in the last photograph is partly due to the watering effect of the sloping wire netting and to the previously mentioned edge effect, but if this luxuriant vegetation were exposed to the rabbit attack it would of course very quickly be eaten down to the low constant level of the very dwarf general vegetation of the exposed grass heath on the distant side of the netting.

The *Carex arenaria* inside the rabbit-proof enclosure is growing more vigorously and taller than the grasses now that it has been protected from the heavy rabbit attack. Vigorous leaf blades of *Carex arenaria* can be seen in Photo. 34 associated with the vigorous flowering young *Calluna* stems. If the vegetation inside the large rabbit-proof enclosure were subjected to a certain small constant intensity of rabbit attack probably the *Carex arenaria* inside the enclosure would eventually dominate the *Calluna*, but as the vegetation inside the enclosure has been *entirely* cut off from the rabbit attack, the *Calluna* inside the enclosure will probably dominate all the

Carex after the latter has dominated the grass heath (see p. 39), and eventually the whole of the enclosure will probably become dominated by healthy *Calluna* heath while the whole of the surrounding area becomes typical pure grass heath.

If the seeds of appropriate trees become deposited amongst the resulting *Calluna* on the protected area these trees would probably eventually dominate the *Calluna* on this area, and if the enclosed area were big enough a small patch of woodland surrounded on all sides by grass heath would ultimately be produced solely owing to the protection of this patch of ground from rabbit attack.

The exact details of the successive developmental phases of the progressive succession on this enclosed protected area inside the large rabbit-proof quadrat, from grass heath to *Calluna* heath and eventually to woodland, owing to the protection of this particular area from rabbit attack, will probably form a very interesting subject of observation.

CHAPTER V.

OBSERVATIONS RELATING TO COMPETITION BETWEEN PLANTS.

ON THE MANNER IN WHICH COMPETITION ACTS BETWEEN PLANTS IN MIXED ASSOCIATIONS.

Amongst the innumerable examples of isolated plants of the same species competing with neighbouring plants of different species in Breckland one of the most interesting is the case of *Juncus squarrosus* which occasionally exists in the valleys and sometimes succeeds in suppressing the neighbouring plants by means of its characteristic method of producing cup-like rosettes of strong stiff leaves which flatten out upon the surrounding vegetation and tend to kill it[1].

Plants which are more dwarf than the *Juncus squarrosus* itself are the ones which tend to suffer most from this method of suppression. *Agrostis alba* is the commonest plant in Breckland to suffer in this way, since it is often prevented by the rabbit attack from growing tall while the rabbits do not eat the tough *Juncus* leaves so readily except when the rabbit attack is very heavy. When however the rabbit attack is very heavy the taller and more vertically growing *Juncus* leaves suffer more severely from the heavy rabbit attack than do the more dwarf growing *Agrostis* leaves. They are then nibbled down closely to the surface of the soil and are unable to suppress the neighbouring *Agrostis* plants.

Sometimes dense groups of *Juncus squarrosus* plants occur, and often in these instances the central plants of a group are all dead and only the *Juncus* plants on the outer edge of the group are alive forming a sort of "Fairy Ring" of *Juncus squarrosus*. In some of these instances the death of the central plants may perhaps be partly due to some of the *Juncus* plants within the

---

[1] **Lindman, C. A. M.**, "Some cases of Plants suppressed by other Plants," *New Phytologist*, **12**, 1913, p. 3.

close-set colony having been suppressed by others, so that only the *Juncus* plants on the extreme edge of the colony remained alive.

The effects of the artificial experimental increase of the available water supply on the result of the competition of the various species of plants in the grass heath association have already been dealt with (see p. 48). The increase in the water supply alone enabled *Agrostis vulgaris* to grow sufficiently luxuriantly to crowd out and smother *Festuca ovina* and all its other competitors which are present under normal conditions in the grass heath association. Etiolated, dying, and dead remains of these various competitors could be found amongst and underneath the tall and luxuriant growth of *Agrostis*. In this instance, alteration in one factor alone of the environment has greatly affected the result of the competition, and has decided which of the competitors shall survive in the struggle for existence.

After having observed the details of the above instance of the way in which competition acts in Breckland, the writer examined numerous pastures with a view to finding other instances of analogous phenomena. Inside the edges of patches of taller growing rhizome-spreading grasses (such as *Poa pratensis*) in pastures, blanched, dying and dead remains of more dwarf growing grasses, and small pale clover leaves on weak etiolated petioles, can often be found, while the lower portions of the stems and leaf blades of the taller growing grasses themselves are often etiolated.

In this connection it may be noted that, as is well known, many plants which in nature are very rare and only occur under very special conditions of soil, etc., can be got to grow well in gardens in very different and various soils and under very different conditions of soil moisture, etc. This is chiefly because in gardens these otherwise rare plants are protected by man from being smothered at some period of their life by various other plants (weeds, etc.), which would otherwise grow taller and more abundantly than they under these different soil conditions and would thus be able to smother them.

These instances illustrate the very great importance of tall and luxuriant growth of the *aerial* portions of plants from the point of view of ultimately successful competition and the extermination of competitors, and the very great importance and significance of this is often insufficiently realised, nor is it sufficiently clearly pointed out or emphasised in existing ecological literature. The fact that so many plants when protected from smothering by other plants can be got to grow, or at least to exist, in such very different and varied soils and under such varied conditions of soil moisture, etc. suggests that in nature different and varied soil conditions, whether dependent on the original nature of the soil or induced by root competition with other plants, have in themselves in many cases little to do with their actual *extermination*. Capability of taller growth than that of its competitors under different conditions of soil and moisture and biotic attack, etc. appears to be usually far more important to a plant for its survival and dominance and extermination of competitors on a particular area than mere differences in the soil conditions and the effects of root competition as such, and variations in the soil conditions probably act chiefly indirectly in limiting the distribution of plants and in exterminating competitors. The effects of root competition and soil conditions in these respects, though they may be very great, are probably chiefly important *because and in so far as they affect the differential capabilities of the aerial portions of various plants to grow taller and more luxuriantly than, and then to smother, those of their competitors,* and *in this manner* to exterminate many different kinds of plants which could otherwise grow on these particular soils.

In other words different soil conditions and the effects of root competition for nutriment greatly affect the absolute rates of growth of the aerial portions of the different competitors, but this mere retardation of growth does not in itself usually result in the death of the competitors. In affecting the absolute rates of growth of the aerial portions, however, the effects of different soil conditions and the effects of root competition thereby greatly affect the relative rates of growth of, or the race between, the aerial portions, and the resulting smothering very frequently terminates in the death of the competitors.

The smothering of the more dwarf *Festuca ovina* plants by the taller growth of *Agrostis vulgaris* owing to the greater capacity of taller growth possessed by the *Agrostis vulgaris* when the soil moisture was increased in the water drip experiment upon the available water supply is very interesting in this connection. No doubt the very small natural available water supply in the soil was also limiting the growth of the *Festuca ovina*, and when this supply in the soil was increased the *Festuca ovina* would doubtless also have been able to grow better than before if it were not smothered under the altered conditions of increased water supply to the roots of both kinds of plants by the *still taller* induced growth of the competing aerial portions of the *Agrostis*. In this instance the fatal result of the competition to the *Festuca* could clearly not have been the effect of root competition for nutriment *per se,* for the available water supply, and also on the manured side the available manurial supply, to the *Festuca* roots as well as to the *Agrostis* roots were very greatly increased from their previous values. It is apparently an individual case of the smothering of a plant by the taller and more luxuriant aerial portions of a competitor.

Probably the survival of many rare plants only under very special conditions in nature is chiefly because only under the special conditions of the habitat (of soil, biotic attack, etc.) can no other plant grow taller and more luxuriantly than they and thus be able to smother and exterminate them. No doubt the usual heavy mortality rate of seedlings is very largely due to their having to compete from the first with previously well-established taller growing competitors.

The view that plants chiefly and usually exterminate their competitors by competition between and smothering by the aerial portions and not by root competition for nutriment *per se* is strongly confirmed by the growth of the various plants on the experimental grass plots at Rothamsted. On the unmanured grass plots sixty or more species of plants occur, but when the supply of nitrogenous and mineral manurial constituents to the roots of all the plants is increased, and acidity is neutralised by lime, the aerial portions of about six species of plants eventually grow so vigorously as to smother practically everything else in spite of the fact that the available supply of food constituents has been increased to the roots of all plants. It is difficult to see how root competition for nutriment *per se* could usually be the effective factor in the extermination of competitors if when the available supply of nutriment to the roots of all plants is increased the number of species present nevertheless falls from sixty to six. The reduction in species consequent upon increased supply of food constituents to the roots of the plants is in some respects contrary to what might have been expected on *a priori* considerations; but the explanation chiefly is that when owing to poverty of soil and root competition for nutriment the aerial portions of no species can grow luxuriantly many species of plants then have a chance of life without being smothered and exterminated by the aerial portions of more luxuriant competitors

The above explanation of the comparative poverty in species of the flora of rich soils that it is chiefly due to aerial smothering is probably the correct explanation of a phenomenon which has long puzzled the writer. The alluvial soil of the Holland Division of South Lincolnshire is very rich and the agriculture is very intensive, a consequence being that waste ground is comparatively small in area, and that woods and commons as ordinarily understood are practically absent. Nevertheless waste ground does occur along roadsides and in various other places, and the flora of this waste ground is very poor in species compared with that of most other parts of England.

No doubt this poverty in species of the waste ground of this district is largely due to the intensive agriculture of the surrounding region and to the absence of extensive woods and commons which could act as perpetual centres for the distribution of various plants; but the relative poverty in species on the existing waste ground seemed far too striking to be put down to this cause alone. If however it is caused by the relative abundance of growth of the aerial portions of a comparatively few species of plants on the rich soil smothering the potential growth of many otherwise possible species the striking poverty in species of the wild flora of this district would be largely explained. In order to estimate the magnitude of the separate effect of the absence of woods and commons it would be interesting to compare the extents of spread of various plants from individual woods and commons on the poorer ground at the edge of the rich alluvial plain on to the rich soil and along the poorer ground respectively. It seems likely that such plants would be found scarcely able to survive or spread effectively at all amongst the more luxuriant and taller vegetation on the waste areas of the rich soil.

It appears probable that plants usually kill their competitors by so reducing the available value of some factor or factors which are limiting the metabolism and growth of their competitors that the individual competitors are eventually killed off by this reduction though they may struggle on for a time by living on their reserves. In this connection the experiments of physiologists in feeding animals on gradually reduced protein rations are interesting. Probably taller growing plants often reduce the light available for their more dwarf competitors to a value considerably below its external value so that light becomes a severe limiting or controlling factor to the life processes and growth of the more dwarf competitors (and the frequency of etiolation among clovers and grasses supports this) and the latter are eventually killed off or rendered liable to fungal attack by the reduction of metabolism and consequent exhaustion. Limitation in the amount of various other factors to the smothered plants at different times may possibly also have a bearing upon the matter. Experiments are badly needed to settle these points.

COMPETITION BETWEEN PURE PLANT ASSOCIATIONS.

Breckland is very rich in examples of almost pure plant associations competing with one another, in addition of course to innumerable examples of the more ordinary conditions in which isolated plants of the same species compete with neighbouring plants of different kinds.

The chief instances of almost pure or relatively pure plant associations competing together in Breckland are represented by the following pairs of dominants—

Calluna vulgaris and *Carex arenaria.*
Pteris aquilina and *Calluna vulgaris.*
Pteris aquilina and *Carex arenaria.*
Calluna vulgaris and *Erica tetralix.*
Juncus effusus and *Salix repens.*

Some of these cases of competition between almost pure plant associations have already been mentioned, and the very important influence which the presence of rabbits exerts upon the competition has been described. (See Chapter III.)

The rabbits by attacking some plants more than others indirectly confer enormous advantages upon the plants which are least attacked—especially upon *Pteris aquilina* which is only slightly attacked—and when some plants can avail themselves fairly rapidly—such as by rhizome growth and reproduction in the case of *Carex arenaria*—of the opportunity given them by the rabbits attacking their competitors, the effect of the rabbits upon the competition becomes very quickly apparent in the resulting rapid movement of the various plant associations.

Although plants which can grow tall normally possess enormous advantages over more dwarf competitors yet it has already been mentioned that those plants which naturally tend to grow tall are usually the ones which eventually suffer most from heavy and increasing biotic attack, largely because upright growing stems attract the attention of the animals and are easier for them to eat, and because the stems of taller growing plants such as seedling trees are usually fewer in number per unit area than those of more dwarf competitors and are in consequence more readily exterminated. (See p. 27.)

It has already been explained that the dominance of the different kinds of vegetation on the ground occupied by their respective zones in the characteristic zoned vegetation around rabbit burrows in Breckland, and also the dominance of the same kinds of vegetation on the extensive areas occupied by the corresponding large *Calluna, Carex,* and grass heath plant associations of Breckland, largely depend upon and represent the result of a dynamic balance between these two strongly opposed tendencies. (See pp. 34–41.)

Thus in Breckland the existence and dominance of the large grass heath, *Carex arenaria* and *Calluna vulgaris* associations on the ground of their respective areas ultimately depend upon the injurious effects of certain intensities of rabbit attack severely injuring the taller members of the series sufficiently to allow the more dwarf kinds of vegetation to become dominant.

In other words a small intensity of rabbit attack which is just sufficient to kill off seedling trees on an area allows the relatively dwarf *Calluna* to become dominant, a somewhat greater intensity of rabbit attack kills off the *Calluna* and allows the more dwarf *Carex arenaria* to dominate the particular area, and a still heavier rabbit attack kills off the *Carex arenaria* and allows the still more dwarf grass heath to dominate the area. If it were not for the more injurious effect of the biotic attack upon the taller plants, these other more dwarf types of vegetation would quickly become smothered and replaced by the next taller member of the series, and they would eventually all become replaced by still taller woodland.

Thus the above mentioned highly characteristic plant associations of Breckland ultimately depend for their distribution and dominance on the ground of their respective areas upon the maintenance of certain particular intensities of rabbit attack over these areas, and it will thus be realised what important effects the rabbits have upon the competition and dominance of the different kinds of plants, and how—when the rabbit attack is varying—these effects result in producing rapid movements in the distributions of the various plant associations owing to upsetting and altering the ultimate effects of the extremely interesting moving balance between the two strongly opposed tendencies of greater susceptibility to increasing biotic attack and the natural advantages of tall growth.

It has already been mentioned that the zonation and distribution of the different types of vegetation on the other uncultivated areas of England which are subjected to biotic attack very likely largely depend upon and represent the effects of a dynamic moving balance between these two strongly opposed influences.

Although the final dominances and the ultimate distributions of the different kinds of plant associations in Breckland depend upon varying intensities of rabbit attack, yet the actual distributions of the various associations at any time do not always correspond with the possible distributions which the rabbit attack would permit owing to some plants not being able to spread so quickly as the greater destruction of their taller competitors by the rabbits. This is seen for instance in those cases where *Calluna* heath is degenerating so rapidly to grass heath that the rhizome spreading *Carex arenaria* cannot keep pace or catch up with its degenerating competitor, so that the normally or developmentally intermediate *Carex arenaria* zone has not been able to reach or cover all its otherwise possible distribution at a given time, and part of its possible ultimate area of distribution is temporarily occupied by grass heath. *Pteris aquilina* in Breckland is nearly always in a similar condition, for it can dominate all its usual competitors in the district, and its actual distribution at any time chiefly depends upon or is controlled by the comparatively slow rate of growth of its rhizomes.

Various experiments have been made in transplanting *Carex* and *Pteris* to various places in different positions in the intermediate zones which they would probably occupy if it were not for the limited rate of growth of their rhizomes in order to see if they can grow and form migration circles in these new positions, and the results so far obtained support the above view[1].

With regard to the cases of almost pure plant associations competing together under different conditions of rabbit attack in different localities in Breckland, it is frequently not known whether the two associations are almost in equilibrium in various positions or whether one of them is spreading at the expense of the other, and if so the rate at which it is spreading.

In order to obtain information upon these points many maps and charts of the distributions of the different plants on the competing zones have been made on recorded dates, such as charts of *Pteris aquilina* competing with *Carex arenaria* and with *Calluna vulgaris,* etc., and many orientated photographs of competing zones have been taken with a view to comparison with subsequent charts and photographs.

---

[1] It was anticipated that there might be some difficulty in getting rhizomes of *Pteris* to grow after transplanting but no such difficulty has occurred in practice. If the original soil around the rhizomes is not disturbed and the transplanting is done carefully, the *Pteris* rhizomes grow quite well in their new positions and produce healthy fronds. It is however usually necessary for the transplanted plants to be protected from rabbit attacks at first, for even though they be normally little attacked by rabbits, yet isolated in a new position they are subjected to a heavier attack per individual frond, and are at first more of a local novelty, and probably on this account they are more subjected to a process of sampling by the rabbits until they become comparatively numerous and well established.

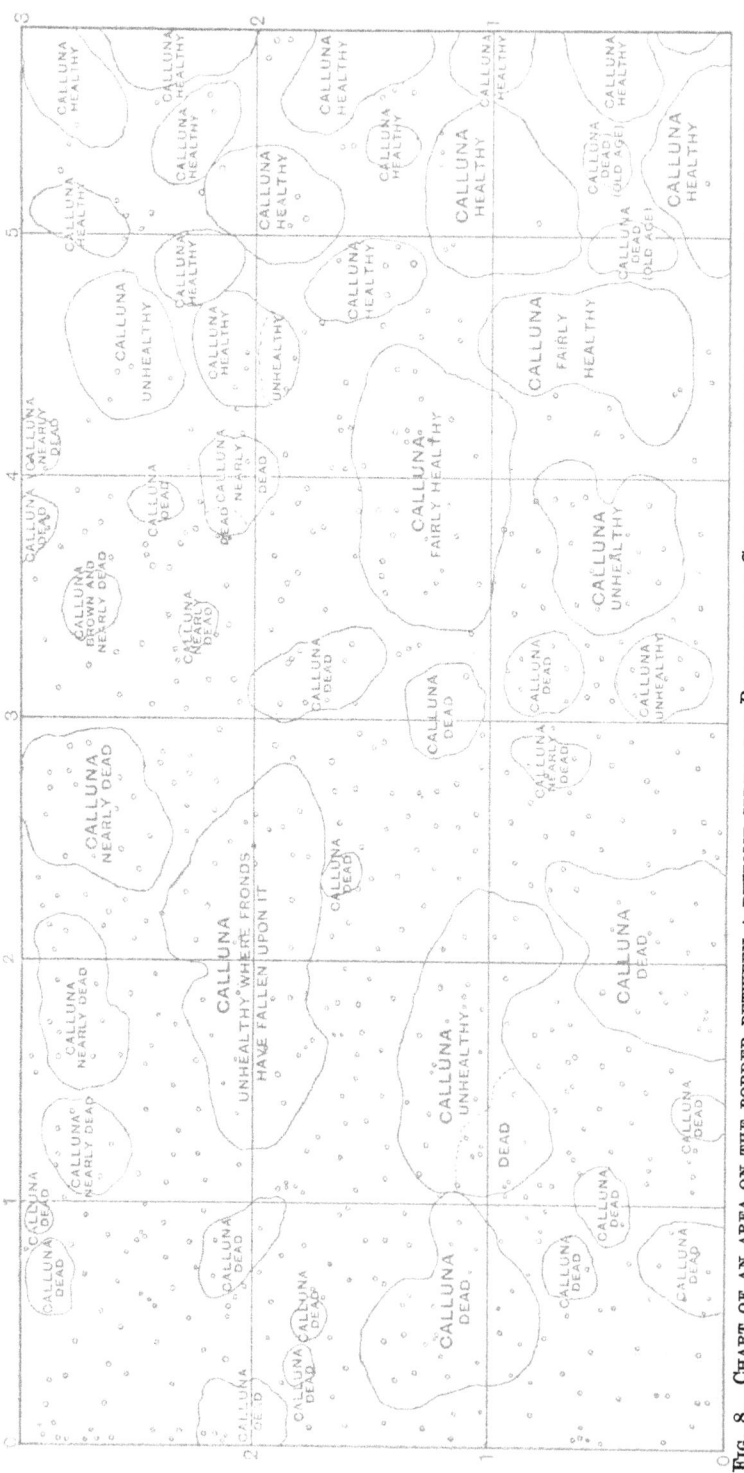

Fig. 8. Chart of an area on the border between a rhizome spreading *Pteris* and a *Calluna* heath association on Cavenham Heath. The *Pteris* fronds are represented by small circles and the *Calluna* bushes by their outlines. The *Pteris* is spreading by rhizome growth from the left-hand side of the diagram towards the right, and many of the *Calluna* bushes on the left-hand side where the *Pteris* has been for some time have been killed by the dead fallen-over *Pteris* fronds while the *Calluna* bushes on the right which the *Pteris* has only recently reached are comparatively healthy.

PTERIS AQUILINA KILLS ITS COMPETITORS CHIEFLY BY MEANS
OF ITS FALLEN-OVER DEAD FRONDS.

When charts of the different kinds of vegetation on competing zones
between *Pteris* and *Calluna* and between *Pteris* and *Carex* associations were
being made it was noticed that the boundaries between the two associations
were usually very narrow, and that dying and dead *Calluna* and *Carex* plants
respectively usually occurred inside the edges of the *Pteris* associations. Hence
it appeared that in most cases the *Pteris* was spreading and was able to kill
the competing *Calluna* and *Carex*. In an endeavour to find out the cause of
this fatal effect of *Pteris* upon *Calluna* and *Carex* the root systems of the competing plants were first carefully examined. The *Pteris* rhizomes run at
an average depth of 25 cms. from the surface and most of their adventitious
roots are still deeper. On the other hand most of the *Calluna* roots occur
between the depths 5—15 cms. from the surface, while the *Carex* rhizomes
run at an average depth of 7 cms.

It will thus be seen that the *Pteris* and *Calluna* and the *Pteris* and
Carex roots respectively occur at very different levels in the soil and that
the root systems of these respective competitors might be said to be com-
plementary[1].

It must be remembered however that though these root systems occupy
different layers in the soil, yet water readily moves downwards and upwards
through the soil by capillarity to any drier zone, so that these root systems
although they occupy different layers in the soil probably compete effectively
together for the available water supply. Not much information about the
manner in which the *Pteris* was able to kill the *Calluna* and the *Carex* was
derived from the examinations of the root systems of the competitors.

Another portion of a narrow transition zone where *Pteris aquilina* was competing with *Calluna vulgaris* and apparently killing it was then carefully charted
in 1913, the positions of all the *Calluna* plants and *Pteris* fronds being carefully
recorded (see Fig. 8). The area was also photographed so that all these records
could be compared with other records made later on and information obtained
relating to the speed with which the *Pteris* could kill the *Calluna*. When
this last mentioned area was being charted the bracken fronds were in the
dead winter condition, and it was noticed that the dead fronds had fallen
over—apparently blown by the wind as they were all in the same direction—
on to the competing *Calluna* bushes (see Pl. XVIII, Photo. 35), and that under
the blown-over dead bracken fronds the *Calluna* leaves were etiolated and often
dead. In these cases where the dead bracken fronds had fallen upon only
part of a *Calluna* bush it was noticed that the leaves on that portion were
usually etiolated and often killed while the portions of the same *Calluna*

---

[1] **Woodhead, T. W.** "Ecology of Woodland Plants in the Neighbourhood of Huddersfield,"
Journ. Linn. Soc. Bot. **37**, 1906.

bush where no fronds had fallen and which were exposed were perfectly healthy. Hence it appeared that the dead blown-over bracken fronds exerted a very important influence in killing the competing *Calluna*. This effect was far more marked than any effect which could be observed of root competition.

As a preliminary to later experimental investigation of the matter it was thought worth while to measure the comparative light intensities in the open and underneath dead fallen *Pteris* fronds near killed *Calluna* bushes by means of a photographic actinometer. Photographic actinometer paper even when orthochromatic is chiefly sensitive to the blue end of the spectrum while photosynthesis depends more upon the red end. In all probability a somewhat greater proportion of the light of the useful red end of the spectrum would pass through the dead brown fronds than of the blue end which latter would be that chiefly measured by the actinometer, yet nevertheless the results obtained by the actinometer might give some idea of the extent to which the dead fallen-over fronds reduced the useful light intensity. In one typical case the average actinometer time under dead fallen-over *Pteris* fronds in the neighbourhood of killed portions of *Calluna* bushes (seen in Pl. XVIII, Photo. 35) was three minutes, against three seconds in the open. In another case under rather dense fallen-over *Pteris* fronds the average actinometer time was eight minutes as compared with five seconds in the open. Thus the light intensity in these two cases was reduced to $\frac{1}{60}$ and $\frac{1}{96}$ respectively, of its value in the open.

It is not necessary to deduce from actinometer readings that the dead fallen-over *Pteris* fronds kill the competing *Calluna* bushes, for the effect of the dead fronds in killing the *Calluna* can be seen directly in those numerous cases where part of a *Calluna* bush has been covered by them and part has remained free.

It is not intended to infer from the above figures that the fallen *Pteris* fronds kill the competing *Calluna* bushes simply by reducing the light intensity—for example the fatal effect of the dead fronds may partly be due to retention of water and resulting decay of the *Calluna* leaves (*Calluna* leaves are delicate and readily decay). It was noticed however that the *Pteris* fronds sometimes killed the *Calluna* leaves before the fronds themselves began to decay and it is therefore not necessary to invoke the action of a toxin produced by decay of the *Pteris* in order to account for the death of the *Calluna*[1].

Though light intensity in the open is not usually the limiting factor to photosynthetic activity, yet when it was reduced to $\frac{1}{60}$—$\frac{1}{96}$ of its previous normal value it would probably become a severe limiting factor to the continued carbon assimilation of the *Calluna* even during the brightest parts

[1] Cf. **Harold Jeffreys,** "On the vegetation of Four Durham Coal Measure Fells." *Journal of Ecology,* **5,** p. 145 (1917).

of the day. Even with a thinner covering of dead *Pteris* fronds and therefore a less extreme reduction of light to the underlying *Calluna* the reduction might well be sufficient to lessen or altogether prevent carbon assimilation by the evergreen *Calluna* during the dull days of winter.

Calluna is well known to be intolerant of shade and it cannot exist where— in woods and elsewhere—the light intensity is reduced to between $\frac{1}{3}$—$\frac{1}{5}$ of its external value. In view of these facts it seems probable that the shading effect is the chief factor in the death of the *Calluna* leaves by the fallen *Pteris* fronds, for, as above mentioned, they can kill the *Calluna* leaves before the fronds themselves begin to decay. *Cutting off of light* and *retention of water* with resulting decay of any delicate leaves with which they may be associated are probably the chief factors in the fatal effects of fallen *Pteris* fronds. The *Calluna* plants are practically always all completely killed between 5 and 10 metres inside the extreme advancing edge of a *Pteris* association.

It has already been mentioned that when the edge of a *Pteris* association is competing with a *Carex arenaria* association dying and dead remains of *Carex* can usually be found as the *Pteris* association is entered and that this indicates that the *Pteris* can usually kill competing *Carex arenaria*. The manner in which the *Pteris* was able to kill the competing *Carex* could not be discovered by an examination of the relations between the root systems of the competitors, but after the effect of the dead fallen *Pteris* fronds in killing the competing *Calluna* bushes had been observed, further observations were made upon the competition of *Pteris* and *Carex* and it was noticed that the effect of *Pteris* in killing the competing *Carex* was also due to the effects of the fallen-over dead fronds upon the *Carex* in the same way as with *Calluna*. Under dead fallen-over *Pteris* fronds the *Carex* plants were etiolated and frequently dead while neighbouring *Carex* plants upon which no fronds had fallen were quite healthy. The fatal effects of the dead fronds are even more marked on the more dwarf *Carex* plants than on the taller *Calluna* bushes.

The fatal effects of the dead fallen-over *Pteris* fronds upon competing *Carex* plants are very marked even where the fronds are fairly wide apart while they are standing upright for, when they fall right over, they may form a covering of dead fronds several fronds thick owing to the fallen individual fronds then overlapping one another although they were fairly wide apart whilst vertical.

Later on when the upright living fronds grow more densely the lower covering of dead fronds becomes of course thicker, but before this time practically all the ground competitors have usually been killed off by the previously fallen fronds. Actinometer readings showed that the reduction in light intensity to the ground vegetation caused by the layer of overlapping dead fallen fronds on the surface is much greater than that caused by the upright living fronds even when these latter are growing fairly thickly.

In addition to having observed the fatal effects of dead fallen-over *Pteris* fronds upon competing *Calluna* bushes at many places in Breckland, the writer has observed the same phenomenon on Wimbledon Common. On Wimbledon Common a *Pteris* frontier—not so sharp and distinct as the Breckland ones—was advancing into *Calluna* bushes. The *Calluna* bushes which the advancing *Pteris* had not yet reached were quite healthy while most of the *Calluna* bushes on the ground which had been occupied by the *Pteris* for some time had been killed by the dead fallen-over fronds. Some small mounds covered by dead *Pteris* fronds were noticed on this area, and when the coverings of dead fronds on these mounds were raised, dead and decaying remains of *Calluna* bushes were found underneath them. Parts of some of the *Calluna* bushes in the transition zone had been covered by dead fallen fronds and the *Calluna* leaves under these were dead while the remaining exposed portions of the same bushes which had not been covered by fronds were quite healthy. Similar cases have been observed on the Greensand at Potton (Bedfordshire); also on Chobham Common (Surrey), Dersingham Common (Norfolk) and Bagshot Heath (Surrey)[1].

The writer has also noticed dead fallen *Pteris* fronds killing *Ulex* bushes near Welwyn and on Hampstead Heath. In Pl. XVIII, Photo. 36 (taken on Hampstead Heath) *Pteris* rhizomes are advancing towards the middle of the *Ulex* bush from the left-hand side of the picture and the fallen fronds are killing the middle portions of the bush which are underneath them, leaving living portions on each side separated by the dead middle portions as can be seen in the photo. There is a very close correspondence between the distribution of the dead *Pteris* fronds on the bush and the death of the *Ulex* leaves and branches, those covered by dead fronds being quite dead while those portions of the bush which have up to the present remained uncovered by dead fronds are perfectly healthy.

Pteris fronds usually grow much higher inside tall bushes in the *Pteris* associations than they do in the open. This can be seen for instance inside numerous *Crataegus monogyna* bushes on an area outside the eastern edge of the valley fen wood marked 8 on the map of Cavenham Heath. Inside individual bushes in the *Pteris* associations the fronds often grow eight feet high, and one instance has been noted of a frond 11 feet (3·3 metres) growing out of the top of an elder bush (*Sambucus nigra*) whilst outside the bushes they are normally only between three and four feet in height.

When dead the tall *Pteris* fronds inside the bushes tend to remain entangled in the branches after the dead fronds on the surrounding open areas have fallen to the ground or been blown away. Sometimes the tall dead fronds inside the bushes remain entangled in the branches for several years and when this happens accumulations of tall dead fronds produced by several successive seasons exist entangled in the branches and these dense accumulations some-

[1] Several friends whom the writer has told of this phenomenon have since observed instances of it in many other parts of England.

PLATE XVIII

Photo. 35. Competition zone between *Calluna vulgaris* and *Pteris aquilina* associations, Cavenham Heath. Note the narrowness of the zone. Dead remains of *Calluna* are found within the edge of the *Pteris* association, which is advancing and killing its competitor chiefly by the dead fronds falling on and smothering the *Calluna* bushes, as can be seen in the picture. (See pp. 61–2.) The parts of the *Calluna* bushes covered by dead fronds are usually dead, while the parts of the same bushes remaining free from the fronds are perfectly healthy.

Photo. 36. Dead fallen *Pteris* fronds killing middle of *Ulex* bush on Hampstead Heath. The rhizomes of *Pteris* are advancing towards the middle of the bush from the left of the picture. The middle of the bush, covered by the dead fronds, is dead: the portions on either side, as can be seen in the picture, are living and healthy. The correspondence between the distribution of dead fronds on the bush and the death of the *Ulex* leaves and branches is very close. (See p. 64.)

times kill the bushes with which they are associated. The writer has observed a striking case of this phenomenon in a hawthorn bush on Wimbledon Common. The ends of the branches outside the accumulation of fronds had produced a few leaves and flowers, but inside the accumulation of dead fronds no such leaves or flowers occurred. Many of the lateral branches had decayed and the main branch was decaying and was weighted down very badly by the heavy accumulation of dead fronds.

This phenomenon of the dead parts of the plant's body killing the competitors of the plant so effectively is one of widespread occurrence and of considerable biological interest. It has probably had a considerable bearing upon the long survival and extensive distribution of *Pteris* and similar ferns in various parts of the world, e.g. *Gleichenia linearis* (*G. dichotoma*) in the Tropics.

It is hoped to make actual experiments upon the effects of the dead fallen-over *Pteris* fronds in killing competitors, and thus probably enabling the *Pteris* to spread more rapidly, by carefully charting and photographing areas on competing zones and then cutting and removing the dead fronds just before they fall over from large 20 metre quadrats on these areas for several successive years while leaving the fallen-over fronds on the remaining portions of the areas.

Juncus effusus and small bushes of *Salix repens* are competing in some of the valleys of Breckland. When the *Juncus* leaves die they normally fall over on to the ground and in these localities there is normally a stratum of dead *Juncus* leaves several inches thick on the surface. Where however a *Salix repens* bush is growing amongst the *Juncus effusus* the dead *Juncus* leaves remain entangled in the *Salix repens* branches and cannot fall on to the ground. In consequence of this there is often a dense accumulation of dead *Juncus* leaves produced by several successive seasons entangled amongst the *Salix repens* branches (cf. the same thing in the case of *Pteris* and *Crataegus*) and these dense accumulations of dead *Juncus* leaves frequently kill the associated *Salix repens* bushes. In the case of more dwarf *Salix repens* bushes the dead *Juncus* leaves kill them by falling right over on top of them, for the contents of small raised heaps in the normally level deposit of dead *Juncus* leaves were examined and dead and decaying remains of small *Salix repens* bushes were found underneath the heaped up *Juncus* leaves. This killing of *Salix repens* bushes by dead fallen-over *Juncus effusus* leaves resembles and reminds one of the killing of competitors by dead fallen-over *Pteris* fronds but it is not of such widespread occurrence. These instances of dead discarded portions of taller plants falling over upon and killing more dwarf plants are instances of some special advantages which certain taller growing plants have over their less tall competitors.

A SINGLE ROW OF PINE TREES ACTS AS A BIOLOGICAL BARRIER TO THE
RHIZOME-SPREAD OF *CAREX ARENARIA* OWING TO THE LAYER
OF DEAD FALLEN PINE LEAVES.

A patch of *Carex arenaria* much eaten down by rabbits occurs on the grass heath association of Cavenham Heath near the spot marked 17 on the map. (See p. 6.) The origin of this isolated patch of *Carex* was a mystery for some time. Its presence was at first usually attributed to some edaphic difference in the grass heath at this spot and it seemed possible that this might be the true explanation and it was thought of making analyses of the soil of this patch. Eventually however the writer suddenly noticed that this large patch of *Carex* on the grass heath was opposite a fairly wide gap in a single row of pine trees which runs from the point marked 11 to near the area marked 4 on the map[1] and was connected with the main *Carex* area on the other side of the row of trees by a neck of *Carex* passing through the gap (see Fig. 9).

It appeared on examination that the *Carex* patch owed its existence to the *Carex* having spread through this gap in the row of pines on to the main degenerate heath association and that the *Carex* had only spread through the gap in the row of trees and had not spread through the row of trees itself. Where the row of trees was fairly continuous the otherwise advancing *Carex* was kept back at a distance of about six metres from the trunks on the other side of the row. This case of the stoppage of the spread of *Carex* by a row of pines is an extremely pretty instance of a biological barrier to the migration of a species. It seemed also that this case of trees stopping the growth of the sedge might perhaps ultimately be found to have some bearing upon the important matter of the relation of a grassland type to forest, and a considerable amount of attention has therefore been given to it.

An area 150 metres long by 75 broad along the row of pines where the *Carex* is usually stopped but passes through the gaps has been carefully mapped and charted on a large scale by the gridiron method (see Fig. 9). Many orientated photographs of the limiting zone have also been taken in order that the area may be recharted and rephotographed at subsequent dates and any movement of the distribution of the *Carex* in relation to the row of pines detected and measured.

It can be seen from the chart (Fig. 9) that opposite narrower gaps the *Carex* approaches the row of pines in a uniform curve but is usually just unable to pass right through the gap. Various hypotheses to explain the stoppage of the advance of *Carex* were examined. It was thought that the stoppage of the *Carex* might possibly be due to the shade cast by the trees, but a comparison of the light reduction by the pines with the distribution of the *Carex* in various places by the help of a photographic actinometer

---

[1] This single row of pine trees, which acts as a biological barrier to the *Carex*, with gaps of various widths in it, can be well seen in the background of Photos 31 and 32 on Plate XVI.

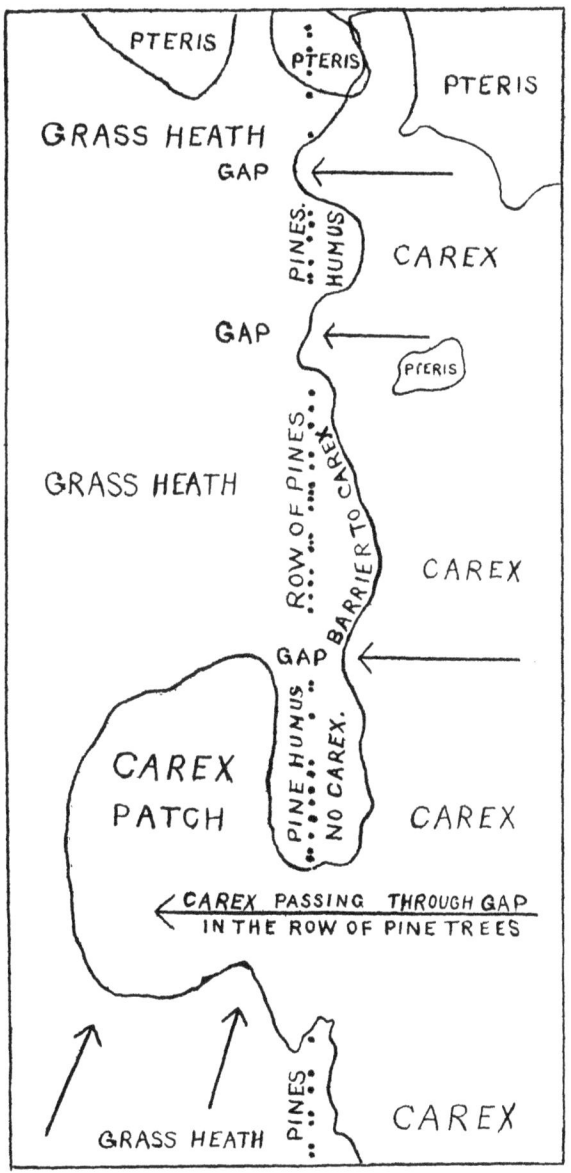

FIG. 9. CHART OF AN AREA ON CAVENHAM HEATH WHERE *CAREX ARENARIA* SPREADING BY
RHIZOME GROWTH PASSES THROUGH WIDE GAPS IN A SINGLE ROW OF PINE TREES ON TO
THE GRASS HEATH BEYOND. Where the row of trees is more continuous the thickness of
the layer of pine litter acts as a definite biological barrier to the rhizome spread of the
Carex (see pages 66–70). Note that the *Carex* approaches or spreads into but not through
the three narrower gaps in the row above the main gap. The *Carex* which has passed
through the main gap in the row of trees forms a patch which is asymmetrical with regard
to the gap. This is owing to the heavy rabbit pressure from the main area of the degenerate
grass heath acting in the direction indicated by the arrows in the lower left-hand corner
and keeping the *Carex* from spreading in that direction.

showed that there was no definite relation between them, and this hypothesis appeared inadequate to account for the observed phenomenon. It was also thought that the advance of the *Carex* might possibly be stopped by the deposit of pine needles and cones dropped by the trees, but a general examination of the distribution of the deposit on the area revealed no clear relation between the distribution of this deposit and the stopping of the *Carex*, very many pine needles and cones occurring actually amongst the bordering *Carex*—nearly as many as on the area under the trees which the *Carex* had failed to occupy.

A careful examination of the distribution of the *Carex* in relation to each of the individual trees in the row was then made, and several instances were discovered in which the *Carex* was absent *vertically* below the branchlets on the extreme ends of certain downwardly inclined pine branches whose tips were comparatively close to the ground; although some *Carex* frequently existed between the completely bare areas under the ends of these branches and the actual trunks of the trees. In such cases the stoppage of the *Carex* seemed almost certainly due to the thicker and greater direct vertical deposit of pine needles and cones dropped down from the comparatively near branches, possibly partly combined with the shade of the branches. The deposit of needles and cones was therefore carefully removed from various of the bare areas under the ends of downwardly inclined branches, as well as from various places on the general limiting zone of the *Carex* along the row of trees, in order to see what was happening to the rhizomes which were endeavouring to invade the bare areas. It was found that as the rhizomes grew further and further under the pine branches the vertically growing *Carex* shoots and leaves became more and more etiolated until ultimately they were very yellow and weak and would not remain vertical without lateral support after the pine humus had been removed. No evidence was found of fungal attack. The *Carex* rhizomes run in the underlying sand, and as the vicinity of the pine branches is approached the deposit of needles and cones above the underlying sand becomes gradually thicker until eventually the deposit is so thick that the *Carex* shoots and leaves springing from the rhizomes cannot reach the free aerial surface, and they are completely etiolated and eventually die. The stoppage of the spread of the sedge is thus seen to be due to the thickness of the deposit of pine needles and cones dropped by the pines.

In this connection it is interesting to note that the *Carex* near the limiting edge typically grows in the form of characteristic vigorous but isolated tufts. This is probably because in this zone a few *Carex* shoots and leaves from the rhizomes in the underlying sand occasionally penetrate the above lying stratum of pine humus in comparatively weak spots resulting in the production of local isolated centres of assimilates and in the production of compact dense tufts of *Carex* shoots and leaves at these spots[1].

[1] This production of dense compact tufts of *Carex* shoots and leaves at isolated spots where the thick deposit of pine humus is locally penetrated strongly reminds one of the production

At the northern end of the chart (Fig. 9) portions of an area are seen where *Pteris aquilina* has spread right under the row of trees. The bracken can spread right up to and through the row of pines even where the trees are close together, while *Carex* cannot. This difference probably chiefly arises because the *Pteris* rhizomes typically occur at a much greater depth (an average of 25 cm. from the surface) than the *Carex* rhizomes, which usually run at a depth of about 9 cms., and therefore the young coiled (circinate) *Pteris* fronds can reach the free aerial surface through a much greater thickness of overlying litter than can the young *Carex* shoots and leaves.

It is intended to experiment on the effect of pine litter on the spread of *Carex* by marking out, charting and photographing two large quadrats on this area along the limiting edge of the *Carex*. A certain depth of pine litter will then be removed from one of these quadrats and deposited evenly over the surface of the other quadrat. After a certain interval of time, the whole area will be recharted and also re-surveyed photographically, and any alteration in the distribution of the *Carex* on the experimental quadrats will be ascertained and measured and compared with any movement in the distribution of the *Carex* on the control areas at the ends.

Since the stopping of spreading *Carex* by a single row of pines was first studied on Cavenham Heath, similar phenomena have been observed in other parts of Breckland. *Carex arenaria* often grows closely around isolated pine trees[1] and sometimes it even grows uniformly throughout pine woods. It frequently happens however that otherwise spreading *Carex* is absent from the area under the branches of single isolated pine trees. Such trees are usually well established and old, and in these cases there is usually a thick deposit of pine litter—sometimes over 20 cms. thick—on the areas bare of *Carex* around the trunks. The *Carex* on the limiting zones frequently grows in

---

of dense compact isolated tufts of vegetation on the heavily manured grass plots at Rothamsted, where the conditions of manuring are such that decomposition is retarded and a dense layer of humus remains permanently on the surface of the soil. This dense layer of humus is only penetrated by plants at isolated spots resulting in the production of the compact isolated tufts of vegetation at these points.

[1] In some few cases *Carex* is more luxuriant immediately under isolated pine trees than it is away from them. In one such case on Tuddenham Heath the maximum length of the *Carex* leaf blades in the open was 40 cms. whilst under some isolated pine trees the leaf blades of the *Carex* attained a maximum length of 70 cms. This is contrary to the usual condition of things, but these special reversed cases were always in valleys, and it appeared very probable that in these places there was rather more water than the *Carex* preferred, but round the pine trees the water would be reduced and the *Carex* could grow better. If this be the true explanation, these special cases of association of *Pinus* with *Carex* might perhaps be termed semi-complementary associations. The pines here help the *Carex* plants on the whole; but probably the presence of the *Carex* does not help the pines but is a disadvantage to them. If this be so these special associations of *Pinus* with *Carex* might perhaps be termed semi-communal associations, for under these conditions the effects of competition between the *Carex* and *Pinus* plants are masked as far as the *Carex* is concerned, and the pines help the *Carex* plants. These particular associations are real semi-"communities" in some respects.

the form of the vigorous isolated tufts already described, and careful examination of the shoots and rhizomes which were endeavouring to penetrate the humus-covered region revealed similar etiolation and ultimate death of the young rising shoots where the overlying layer of pine humus had exceeded a certain thickness. The stoppage of the *Carex* in these instances is thus also apparently due to the thick deposit of pine litter.

It is interesting to note that the *Carex* in Breckland can usually grow quite close up to and around the trunks of most broad-leaved trees such as the oak. The writer was at first inclined to believe that this particular difference might be caused by some toxin given out by the conifer humus, but is now inclined to believe that it is primarily and chiefly if not solely due to the conifer humus acting as a far more compact and thicker mechanical obstacle to any young rising shoots which endeavour to penetrate it. The narrow pine needles and heavy cones drop comparatively vertically when they fall from the tree branches and thus they form far more definite surface accumulations of humus below the tree branches, settle down comparatively closely to the surface of the soil, are not so liable to be blown away, and form a dense compact mat around any young rising shoots at the surface of the soil. Also pine needles are not converted into humus so quickly as the leaves of most broad-leaved trees and thus they have time to form far deeper mats or accumulations on the surface. The writer is thus of opinion that it is the sheer mechanical obstacle of the thick compact, dense, impenetrable surface mat of closely packed dead needles and cones which probably chiefly causes the observed absence of vegetation under the narrow-leaved conifers. The fact that the *Carex* grows luxuriantly in the presence of much pine humus provided the latter does not exceed a certain depth, and the fact that it grows in the form of *vigorous* tufts where it has succeeded in locally penetrating an *even thicker* layer of pine humus, seem to indicate that the effect of the pine humus is simply mechanical and to exclude the existence of a definite toxic effect—at least in the case of *Carex arenaria*[1].

[1] Cf. **Harold Jeffreys**, "On the Vegetation of Four Durham Coal Measure Fells," *Journal of Ecology*, **5**, pp 152–154 and Plate XIX.

PLATE XIX

Photo. 37. Small bare sandy area on Cavenham Heath, associated with convex tufts of vegetation. The origin of these bare areas was long mysterious (see p. 71). It was eventually discovered that they are caused by sand blasts, aided probably by rabbits scratching and loosening the surface sand in the first instance. Note the large quantity of rabbit dung.

Photo. 38. Mushroom-shaped cupola (sand-blasted hummock) on Cavenham Heath. The sand blast has strongly undercut the edge of this hummock, exposing roots and producing a mushroom-shaped structure. These hummocks rise relatively to the surface of the soil largely by the wholesale cutting away and lowering of the latter. Note the large number of flints and fragments of *Calluna* roots and stems exposed on the surface of the soil. Eventually the cupola is completely severed and the loose pellets thus formed are often blown considerable distances.

CHAPTER VI.

CHARACTERISTIC BARE AREAS AND SAND HUMMOCKS.

SYNOPSIS.

Characteristic areas bare of vegetation and curious loose pellets of turf occur on some of the grass heath areas of Breckland and also in the degenerating *Calluna* zones. The origin of these was very mysterious and puzzling for a long time. The bare areas are of very variable size, some being very small while others are much larger. The exposed sand of these bare areas is often dark in colour and the bare areas are usually at a slightly lower level than the surrounding areas.

It was at one time thought that the bareness of some of these areas might possibly be due to acid surface drainage soaking into them and killing the associated vegetation. But though this hypothesis appeared to explain the observed facts fairly well in the case of some of the bare patches, yet there were others which it did not explain at all well. Some of these were areas generally bare but possessing characteristic isolated convex patches of vegetation (see Photo. 37, Plate XIX).

THE EFFECTS OF SAND BLASTS.

Eventually however an area was discovered, bare generally, but possessing isolated areas of vegetation which were raised considerably above the general level of the bare area and the upper edges of which were undercut—roots being exposed—strongly suggesting the existence of a surface sand blast (see Photo. 38, Plate XIX).

This particular area was therefore carefully examined during wind and the hypothetical surface sand blast was found to have a very real existence. The sand blast was especially apparent under a lens and when magnified the moving sand grains could occasionally be seen to tear a sand grain away from the isolated raised sandy masses.

In the advanced stages of these bare areas many flints are typically exposed on the surface owing to the sand which was once intermingled with them having been cut away and removed by the wind and by the sand blast of grains already moving.

FORMATION OF HUMMOCKS OR CUPOLAS AND OF LOOSE PELLETS.

In the first stages when the sand blast on the surface is slight the vegetation of small isolated areas is often able to cope with it more or less and these form slightly convex raised areas which gradually become more and more convex owing to the deposition of sand from the general surface upon them (see Photo. 37, Plate XIX). In this stage some of these convex raised areas resemble the cupolas of forts and possibly they tend to take on this characteristic and particular form owing to the bombardment to which they are subjected by the rapidly moving sand grains.

As the areas of sand blast around these cupolas become still more active the cupolas become more and more raised in section, the shoots of the plants rising up through the deposited material until the kinetic energy of the moving particles in association with the critical angle of the material prevents the sectional curves from becoming any more convex, the new moving material slipping down the sides of the cupolas and tending to remove some of the already deposited material.

After this the continued sand blast continues to cut away and lower the general surface of the substratum and leaves the now very convex cupolas, which were formerly closed on the general surface, standing on short columns of sand above the general surface of the substratum, chiefly owing to the roots beneath the cupolas tending to hold the sand grains together for a time and to prevent them from being cut away by the surface sand blast[1].

It will be seen that these particular vegetated sand cupolas of Breckland rise relatively to the surface of the substratum in a somewhat different manner from many sand hummocks, for the cupolas arise at first from the surface owing to sand being transported from the general surface immediately around and deposited on them among the plant shoots, so that while the general surface sinks the hummocks rise. Later on this process becomes much more pronounced as the general surface layers of the underlying substratum are cut away wholesale by the sand blast, leaving the hummocks standing relatively much higher than the new surface of the substratum on short columns of sand. Thus the Breckland sand cupolas rise relatively to the surface of the substratum largely by the wholesale cutting away of the substratum in addition to the deposition of sand on the tops of the cupolas themselves.

The Breckland sand cupolas may be compared with ordinary sand hummock plants at Blakeney on stable shingle banks on the landward side of

[1] The extent to which typical sand cupolas of Breckland have become better developed during an interval of one year can be seen by a comparison of Photos 39 and 40.

the main dune ridges. Sand is carried by wind from these dunes and from the foreshore and is deposited amongst these plants, and thus the hummocks arise. But the Blakeney sand hummocks do not rise relatively to the general level of the substratum through wholesale cutting away of the latter as happens in the case of the Breckland sand hummocks.

Though the roots below sand cupolas hold the immediately underlying sand together for some time after the general surface of the surrounding sand has been cut away and lowered by the sand blast, yet when the general surrounding surface is sufficiently lowered, the surface sand blast eventually undercuts the edges of the cupolas beneath many of the roots (see Photo. 38,) and this undercutting often proceeds until the upper portions of the cupolas and the roots underneath them are completely cut off by the surface sand blast from the substratum, and these cut-off cupolas with roots are often blown by the wind to considerable distances. This is the origin of those curious and characteristic loose pellets of turf which occur in various places and which had previously been so mysterious and puzzling.

RETROGRESSION OF THE VEGETATION ON CUPOLAS.

The structural changes in the formation and undercutting of the cupolas are reflected in changes in their vegetation.

One of the most important changes in the vegetation of the cupolas as they become more convex and rise relatively to the substratum is the following: *Agrostis vulgaris,* which was previously co-dominant with *Festuca ovina* before the cupolas began to rise, gradually loses its co-dominance and *Festuca ovina* becomes the sole dominant. This change in the vegetation of a rising cupola is of considerable interest in view of the fact that when the ordinary grass heath was experimentally irrigated the reverse change took place, *Festuca ovina* losing its co-dominance and *Agrostis vulgaris* becoming the sole dominant[1]. The result of the irrigation experiment may rather tend to suggest that this change in the vegetation as a cupola rises, *Agrostis vulgaris* losing its co-dominance and *Festuca ovina* becoming the sole dominant, is very likely largely due to the hummock becoming gradually drier as it rises and becomes undercut by the sand blast, since the change in the vegetation is the reverse of that which occurs in the vegetation when the ordinary grass-heath is experimentally irrigated.

As the small hummocks gradually become taller and undercut from the substratum owing to the sand blast across the surface of this underlying stratum, various mosses (principally *Campylopus flexuosus* and *Ceratadon purpureus*) become dominant on the central portion of the upper surface of the hummocks, and later on, when the hummocks are just about severed from the substratum, various lichens (principally *Cladonia coccifera, Cl. cervicornis, Cl. uncialis* and *Cetraria aculeata*) usually become established on the tops

---

[1] See Chapter IV.

of the hummocks, a ring of just living *Festuca ovina* frequently remaining round the edge of the cupolas, so that the same cupola frequently exhibits zoned vegetation corresponding to the last three stages of this retrogressive succession on its surface (see Photo. 39, Plate XX), the various zones gradually expanding outwards as the cupola becomes gradually drier.

It will be seen that the stages of this succession are as follows:

1. *Agrostis vulgaris* and *Festuca ovina* co-dominant.
2. *Festuca ovina*, sole dominant.
3. Moss stage (*Campylopus flexuosus, Ceratadon purpureus*, etc.).
4. Lichen stage (*Cladonia coccifera, Cl. cervicornis, Cl. uncialis* and *Cetraria aculeata*).

This is a true retrogressive succession—probably chiefly in relation to water supply—and it is a very good case, as it is comparatively rare to have a true retrogressive succession passing through four stages.

Sometimes these nearly severed hummocks topple over on one side before being blown away by the wind, and when this happens any ring of living *Festuca ovina* around the edge of the cupola comes into contact with the sandy substratum and, since the hummock usually falls on the lee side, *Festuca* often roots and grows in the underlying sand. Wind blown sand is frequently deposited amongst this newly rooted *Festuca* on the sheltered side of a toppled over hummock, producing a miniature *Festuca ovina* dune on the leeward side (cf. the deposition of sand on the leeward side of developing sand-dunes). Soon however the fallen over hummock is usually either blown away or gradually disintegrated *in situ* by the sand blast and the newly rooted *Festuca* on the surface of the substratum is then exposed to the full force of the sand blast which quickly kills it.

RELICS OF *CALLUNA* IN CUPOLAS.

In the case of fairly large cupolas, the sand blast usually cuts away and disintegrates much of the cupola itself before finally undercutting the remains (this is well illustrated in Photos 40 and 41). In the case of the formerly larger cupola seen in Photo. 40 the disintegrating and tearing effect of the sand blast on the cupola itself before it finally undercuts the remains can be well seen on the right hand side of the cupola where the vegetation is being gradually destroyed and much loose sand is exposed. The sand blast has already cut away a considerable portion of this cupola and has exposed the root of a degenerated *Calluna* plant which still remains fixed in the substratum in its vertical position at some little distance from the remains of the disintegrating cupola. This plant was once contained inside the formerly larger grass heath (degenerated *Calluna* heath) cupola before the sand around it had been torn away by the sand blast.

Sometimes many portions of the roots of the *Calluna* plants which once occupied these particular grass heath areas, before the original *Calluna* heath

PLATE XX

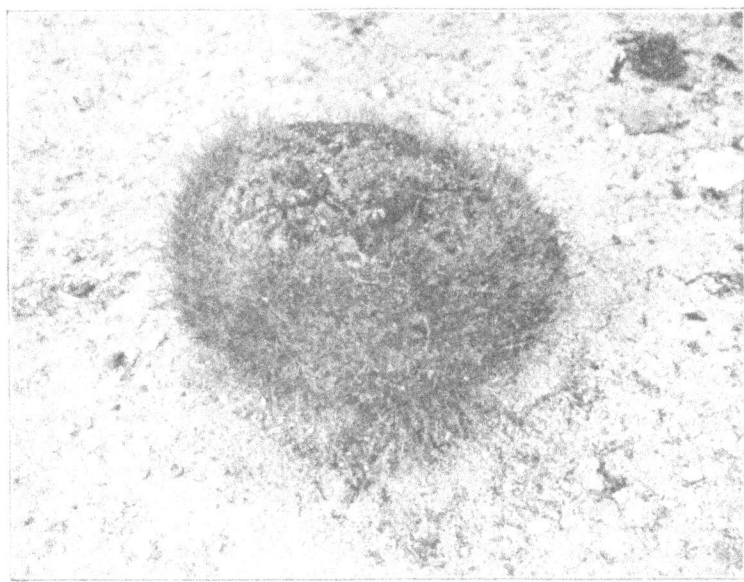

Photo. 39. Zonation of vegetation on a Breckland sand cupola. As a cupola is undercut a true retrogressive succession takes place in its vegetation probably owing to diminution in available water supply. In the case shown a ring of *Festuca ovina* remains round the edge, just above is a moss zone occupied chiefly by *Campylopus flexuosus* and *Ceratodon purpureus*, while the top of the hummock is a lichen zone occupied by *Cladonia uncialis* and *Cetraria aculeata*. (See page 73.)

Photo. 40. Root of a degenerate *Calluna* plant exposed through partial disintegration by the sand blast of a once larger cupola. The sand blast is here tearing away the cupola itself before finally undercutting the remains. On the right of the hummock a degenerate *Calluna* plant with bare roots can be seen. This once formed part of the hummock. (See p. 74.)

degenerated to grass heath through rabbit attack, can be seen remaining fixed *in situ* in the lower portions of isolated grass heath cupolas. In such cases owing to the presence of the numerous tough *Calluna* roots below the cupolas the sand blast cannot undercut the cupolas so readily as it usually can and often begins to disintegrate the upper portions of the cupolas before finally disintegrating or undercutting the remains.

LARGER HUMMOCKS AND SECONDARY HUMMOCKS.

Probably owing to local variations in intensity of sand blast and to other local conditions, considerably larger and taller hummocks are sometimes formed. The vegetation on the upper surfaces of these larger hummocks also passes through the retrogressive succession which has already been described, and eventually when the upper surfaces of these taller hummocks are very dry and rather bare of vegetation the surface layers usually become gradually torn away by the wind and sand blast. This process of the gradual disintegration of the dry upper surfaces of the taller variety of hummocks also frequently exposes fragments of *Calluna* roots, which are all that remain of the former *Calluna* heath that previously occupied these isolated remnants of the higher grass heath surface.

It has already been mentioned that as the smaller varieties of hummocks gradually become undercut the *Festuca ovina* on the tops of the hummocks gradually dies and becomes replaced by various mosses (*Campylopus flexuosus, Ceratadon purpureus,* etc.) and eventually largely by lichens (*Cladonia coccifera, Cl. cervicornis, Cl. uncialis* and *Cetraria aculeata,* etc.). Sometimes in these cases, when the upper surfaces of the primary hummocks are very dry owing to the hummocks being badly undercut by the sand blast across the surface of the substratum, these dry upper surfaces also become sand blasted so that bare sand becomes exposed, and small secondary moss and lichen hummocks filled with sand may form on the upper surfaces of the primary hummocks (see Photo. 41, Plate XXI). Sometimes these smaller secondary moss and lichen hummocks on the dry upper surfaces of badly undercut primary hummocks also have their edges undercut by the sand blast across the upper surfaces. (See Photo. 41.)

RECOLONISATION.

Sometimes, probably owing to unexplained variations in the local conditions, the sand blast across a sand blasted area ceases, and when this happens the cupolas on that area remain fixed and dormant and the intervening bare areas become recolonised. The chief agents in this colonisation of bare sand blasted areas are: *Polytrichum piliferum, Campylopus flexuosus, Cladonia coccifera, Cl. cervicornis, Cl. uncialis* and *Cetraria aculeata.*

Breaking of recolonised surface.

Often the surface layers of these formerly sand blasted areas which have become colonised by mosses and lichens break up into small pieces and these small pieces are blown away by the wind, thus exposing a fresh surface which may again become colonised. A good example of a formerly sand blasted bare area on which the sand blast has become dormant and the surface of which has become colonised by *Campylopus flexuosus* and which has subsequently become broken up into small pieces is seen in Photo. 42 (Plate XXI). Seven ancient formerly sand blasted (now fixed) cupolas are seen in this photograph.

Before their origin and history were discovered, these particular local broken up mossy areas—so different from the surrounding vegetation—were very mysterious phenomena.

Causes of surface cracking and formation of surface scales.

This process of breaking up of the upper strata and scale formation on colonised surfaces is very widespread and of considerable interest. It occurs at Blakeney. The causes of it are not definitely known, but it usually occurs in the surface layers where there are many fibres of the colonising plants and the presence of these fibres would tend to enable the surface layers to withstand tensile stresses in addition to being able to withstand compression stresses, while the lower layers devoid of fibres would chiefly be able to withstand compression stresses alone (cf. reinforced and ordinary concrete). It appears probable that this breaking off of the surface layer from the lower layers has to do with its different capacity for withstanding tensile stresses. This view may be explained as follows:

When the surface soil which previously occupied a certain volume contracts to a smaller volume owing to drying it frequently cracks along lines of weakness where the total tensile stress is greater than the tensile strength of the material. When this happens the upper strata of the soil break up into blocks separated by irregular fissures on the surface but remaining connected below. This is a common phenomenon in dry weather in clay soils and in other soils which—owing to the possession of a certain number of root fibres or other causes—can withstand a certain amount of tensile stresses without breaking down into small fragments. This phenomenon is often a nuisance in agriculture, because mustard (*Brassica nigra, B. alba*), charlock (*B. arvensis*), and other oily seeds, fall down the fissures, and do not germinate in autumn in time to be killed by autumn cultivation, but remain dormant sometimes for many years, and then germinate and grow luxuriantly as weeds when the land is deeply ploughed.

With regard to the formation of loose surface scales on colonised surfaces after some of the lower strata have broken up into vertical blocks, or even in some cases while the lower strata remain longitudinally connected (as in the case of the lower sandy soil in Photo. 42, which cannot withstand tensile stresses without breaking down into minute fragments), the upper surface stratum of the soil contains many fibres of the colonising plants running in various directions. These fibres would enable this surface stratum to withstand a certain amount of internal tensile stress and to contract and expand as a unit within certain limits. The upper layers of this surface stratum are dried by sun and wind and contract more rapidly and before the lower layers. Owing however to the existence of the fibres of the colonising plants running in all directions between these limits the lowest layers of the surface stratum are forced to contract by the tensile stresses at the average rate of contraction of the unit, i.e. at the normal rate of contraction of a layer considerably above them.

PLATE XXI

Photo. 41. Undercut grass-heath hummocks bearing secondary moss and lichen hummocks on their sand-blasted upper surfaces. As a result of the desiccation of a primary hummock by the undercutting of its edges, the upper surface itself often becomes sand blasted and small secondary moss and lichen hummocks may then arise on this. On the left-hand primary hummock the secondary hummocks are also seen to be undercut. (See p. 75.)

Photo. 42. Ancient sand-blasted area colonised by *Campylopus flexuosus* and subsequently broken into small pieces. When the sand blast across an area ceases the area becomes colonised by various mosses and lichens and is often afterwards broken up into small pieces (see p. 75). Seven old, now dormant, cupolas are seen in the photograph.

The upper layer of the substratum, however, which lies immediately below the lower layers of the surface fibre-containing stratum, does not contain many fibres itself and so it can contract at the normal rate of contraction of a layer at its level. Since however the lower layers of the surface stratum lying immediately above it are forced by the possession of fibres and greater average contraction of the surface stratum to contract at the average rate of contraction of the upper surface stratum, i.e. much more rapidly than the upper layer of the substratum, a considerable transverse or shearing stress is produced between these two layers, and, since there are relatively few fibres connecting them, the lowest layer of the surface stratum may be sheared off from the substratum and thus produce loose surface scales. This is the probable explanation of the common phenomenon of the production of loose surface scales on various soils in dry weather.

Later on the upper layers of the loose surface stratum may become considerably drier and tend to contract far more than the lower layers of this loose stratum, with the result that the tensile stresses along the upper zone of the surface stratum become considerably greater than the tensile stresses along the lower zone. When this difference in the longitudinal stresses becomes sufficiently great the loose surface scale may eventually curl upwards especially at its edges. This phenomenon of the curling up of the loose surface scales is very common in various places—for instance on some of the mud flats at Blakeney —especially on those which are only occasionally covered by the tides.

BARE AREAS BETWEEN RABBIT-EATEN *CALLUNA* BUSHES.

In addition to the characteristic bare areas which occur on the grass heaths and the origin of which was puzzling but has now been explained, bare areas sometimes occur between rabbit-eaten *Calluna* bushes (see Photo. 43). The origin of these other bare areas had also been mysterious and various abortive hypotheses had been invented to account for them, but when it was discovered that the bare areas on the grass heaths were due to sand blasts, it was thought that the bare areas which sometimes occur between rabbit-eaten *Calluna* bushes might also be due to sand blasts. If this were so, small cupolas like those found on the bare areas on the grass heaths might also be expected to occur on the bare areas between rabbit-eaten *Calluna* bushes, and in certain places small cupolas were found (see Photo. 43) although they had not been seen before a special search was made. The discovery of these characteristic cupolas on the bare areas between *Calluna* bushes indicated that the bareness in these instances was also due to a sand blast, and on examination during wind the existence of a sand blast was demonstrated. The cupolas are not common on the bare places between *Calluna* bushes—apparently because the sand blasting action has usually gone so far that all the cupolas have been cut off from the surface and blown away.

The past history of these particular areas is probably as follows. These areas were once typical *Calluna* heath but the *Calluna* gradually degenerated through rabbit attack in the way that has been described and grass heath came to occupy the spaces between the *Calluna* bushes. As the bushes became more and more eaten down by the rabbits, the wind would eventually reach the surface of the intervening spaces and start the sand blast, probably in the first instance with the help of rabbits scratching the surface sand and loosening some of it. From a mathematical point of view it is difficult to see how wind alone could raise a sand grain into the air. A loose grain could however be driven up and along a surface by the wind, and on hitting against stationary grains would not only start them moving but might itself shoot up at an angle. When the wind starts a sand grain moving this hits against other sand grains on the surface and starts these moving, and these in their turn start other grains moving and so on. Thus when once the sand blast has started its area tends to spread indefinitely.

The sand blasts which have become started amongst these eaten down *Calluna* hummocks have thus become wider and wider and have cut away most of the grass heath surface which had previously come to occupy the spaces between the eaten down bushes until nothing is left of it except isolated cupolas and hummocks. It would probably never have been suspected that grass heath had ever occupied the now bare intervening spaces between these rabbit-eaten *Calluna* bushes if it were not for these traces of the previous grass heath surface which are left behind in some cases although they have usually all been cut off by the sand blast and blown away by the wind.

Some of the sand which is carried away from the surfaces of the intervening spaces is deposited inside the neighbouring *Calluna* bushes, which are often full of sand. The rabbits usually eventually nibble the stems of the *Calluna* bushes down close to the level of the contained sand, producing very smooth rounded hummocks such as those seen in Photo. 43, Plate XXI. The smoothness of the surfaces of these hummocks is chiefly due to eating down of the *Calluna* stems by the rabbits nearly to the level of the contained sand and is not due to the sand blast as such, for the hummocks are practically as smooth on the side exposed to the prevailing wind and sand blast as on the sheltered side.

Sometimes in the case of very small *Calluna* hummocks which have been eaten down very closely by the rabbits all the contained sand is eventually swept out by the wind, exposing the bare *Calluna* stems.

Surface sand blasts and their resulting effects on the associated vegetation are of fairly widespread occurrence in Breckland. Probably as *Carex* and *Pteris* spread over the areas these plants will largely prevent sand blasts by binding the sand together and sheltering it from wind; but *Carex arenaria* apparently cannot prevent sand blasts in all cases, for sometimes its rhizomes are laid bare by them.

Sand blasted areas on West Newton Heath.

Fairly large sand blasted areas also occur on some of the north-west Norfolk heaths. Such an area on West Newton Heath near Sandringham is shown in Photo. 44, Plate XXI. This heath is largely an *Erica tetralix* heath and is much damper than most of the Breckland heaths; but probably for the existence of a sand blast it is only necessary that the sand grains lying immediately on the surface should be temporarily desiccated. The three

PLATE XXII

Photo. 43. Bare areas between rabbit-grazed *Calluna* hummocks. The bareness of the ground between the hummocks is here also due to sand blast. Small grass heath cupolas like those figured in Plate XIX are seen in the foreground. The whole bare area between the *Calluna* hummocks was probably once covered with grass heath arising from *Calluna* heath through rabbit attack. (See p. 77.)

Photo. 44. Sand-blasted area on West Newton Heath, North-West Norfolk. Three closely rabbit-grazed hummocks of *Erica tetralix*, filled with wind-blown sand, are seen in the middle of the bare area. By attacking the *Erica* on the edge of the bare area the rabbits probably lead to the extension of the area. (See p. 79.)

smooth and rounded *Erica tetralix* hummocks, filled with wind blown sand from the surrounding sand blasted area and closely eaten down by rabbits, on the bare area in Photo. 44 strongly resemble the smooth and rounded rabbit-grazed *Calluna* hummocks of Breckland, filled with wind blown sand, such as those seen in Photo. 43. The curious small tufts of taller flowering *Erica* stems on the left hand side of the otherwise smooth rabbit-grazed central *Erica tetralix* hummock seen in Photo. 44 are probably caused by the rabbits preferring to continue to eat the young terminal growing portions of the already severely eaten down *Erica* shoots rather than the older and tougher portions of taller shoots which have, so to speak, got out of control. The *Erica tetralix* plants around the edge of the sand blasted area in the foreground and on the right hand side of Photo. 44 are closely eaten down by the rabbits and filled with wind blown sand, whilst the *Erica* further away from the bare area is not closely eaten down. Probably the rabbits can reach the *Erica* on the edge of the bared area more readily than the *Erica* which is further away from the edge. If this be the case it reminds one of the greater degeneration of *Calluna* heath from rabbit attack along trackways than elsewhere[1]. Probably the rabbits by eating the *Erica* on the edge tend to help or enable the sand blasted bared area to increase in size, even if they did not originate it.

An extensive area of *Erica tetralix* on West Newton Heath can be seen in Photo. 44. It has already been mentioned that although there are far fewer rabbits on this heath than on most of the Breckland heaths yet owing to the fact that much of the vegetation now-a-days consists of the unpalatable *Erica tetralix* the actual rabbit *pressure* on the vegetation is far greater than that on most of the Breckland heaths, and any more attractive associates of these unpalatable plants would quickly be exterminated by the rabbits. Thus the extensive distribution and the purity of the unattractive *Erica tetralix* association on West Newton Heath are probably largely due to the cumulative effect of the rabbit pressure in killing off any more attractive competitors of these unpalatable plants more and more rapidly as they became rare until they were eventually exterminated[2].

In addition to occurring in Breckland and on the north-west Norfolk heaths bare areas resulting from sand blasts occur on some of the Surrey heaths near Blackheath and on various other sandy English heaths.

[1] See Chapter II, p. 24.   [2] Chapter III, p. 42.

CHAPTER VII.

GENERAL EFFECTS OF BLOWING SAND
UPON THE VEGETATION.

WHEN the effects of small sand blasts upon the vegetation were discovered observation was kept for any other effects which blowing sand might have upon the vegetation.

The blowing of sand seems to be a general phenomenon in many parts of Breckland. It has already been mentioned that when once a sand blasted area is started it tends to spread. Apparently the bombardment of the sand grains which are already moving tends to start other previously stationary sand grains moving and these in their turn tend to start still others, and so on, until a large sand blasted area may be produced.

In the case of these large sand blasted areas there is frequently a cliff of sand at the edge of the denuded sand blasted area furthest from and facing the prevailing wind and sand blast. (See Photos 45 and 46, Plate XXIII.) Beyond the edge of the cliff the higher surface is occupied by vegetation, and the soil consists of tougher strata bound together by plant remains and not yet disintegrated by the sand blast.

Apparently the wind and sand blast on the surface tend to cut away the upper strata of the sand more and more as the length of sand blasted area exposed to the prevailing wind increases. This sand is transported beyond the furthest edge of the sand blasted area, and piled up in this position in the form of a new cliff or dune the face of which in its turn gradually becomes disintegrated by the continued sand blast. The sand blasted area thus gradually increases in length in the direction of the prevailing winds as the face of the cliff is progressively disintegrated.

The accumulated moving sand grains from the increasing depth of sand blasted surface in front of the cliff strike against the face of the cliff and gradually cut it away. (See Photos 45 and 46.)

The height of the cliff above the adjoining sand blasted surface gradually tends to increase both by the increasing removal of the surface strata of the adjoining surface by the sand blast and by the deposition of sand carried by the wind and sand blast on to the upper surface of the bordering cliff. Later on the sand cliff becomes sufficiently high to check the sand from being carried by the wind and sand blast up its face and deposited on the upper surface beyond, and after this the face of the sand cliff often does not increase greatly in height above the surface of the adjoining area. The wind transported sand then accumulates at the base of the cliff instead of being carried up to and deposited on the upper surface beyond. Often however this loose

PLATE XXIII

Photo. 45. Undercut face of a sand blasted Breckland sand cliff on the furthest edge of a sand blasted area facing the prevailing winds. Note the undercutting caused by the sand blast across the lower surface. When this undercutting has procceded sufficiently portions break away and fall to the lower surface where the debris are ultimately disintegrated by the sand blast. Note fragments of *Calluna* etc. produced in this way on the lower surface. Note sand on the upper surface carried there by the sand blast. Many plants can rise through this deposited sand, thus reoccupying the new surface as indicated by the dark streaks in the Photo. (see p. 87). For the bearing of this on Richthofen's theory of the origin of Loess see p. 88. Note the remains of the original layer of vegetation submerged by the deposited sand and now exposed on the disintegrating face of the cliff about six inches below the present surface : also the blow-outs in the face of the cliff on the extreme left.

Photo. 46. Large rabbit-grazed *Calluna* hummock bending over the retreating face of a sand blasted sand cliff as the latter gradually becomes disintegrated. Note the fissures produced in the surface of the hummock as it bends over the disintegrating and retreating face of the cliff (see p. 82). Eventually the hummock will fall upside down on the sand blasted lower surface and will become disintegrated (see p. 82). Note that the lower portions of this cliff project further and resist disintegration better than the upper portions. This is because the lower portions represent the original surface strata long occupied by vegetation and cemented together by humous products, whilst the upper portions represent wind blown sand comparatively recently deposited which is looser and more readily torn away by the sand blast.

transported sand which would otherwise tend to accumulate at the base of tall sand cliffs is removed from this position by winds different in direction from the prevailing winds but not in themselves sufficiently powerful to remove new sand or originate strong sand blasts and new sand cliffs. Some of this loose sand is subsequently again dashed against the face of the cliff by the prevailing winds and some of it is carried to the upper surface.

Thus in the case of long sand cliffs when the wind strikes them at an angle, and in the case of isolated portions of sand cliffs when sand can escape round their edges, transported sand does not accumulate extensively at their bases and especially in these cases the faces of the sand cliffs often eventually come to project a considerable height above the adjoining surface of the neighbouring sand blasted areas owing to the greater and greater removal of the surface strata by the sand blasts, and owing to the deposition of some of the wind borne sand on their upper surfaces. (See Photo. 46.)

A gradual upward sloping tendency of the ground in relation to the prevailing wind probably assists in the formation of sand cliffs. This view is supported by the fact that when the retreating face of a formerly fairly tall sand cliff reaches the downward sloping edge of a slight valley and retreats down the sloping valley side the cliff often gradually becomes obliterated.

Especially in the cases of fairly tall cliffs, their faces generally become undercut by the sand blast. This, no doubt, is largely owing to the roots of the vegetation which occupies the upper surface of the cliff tending to hold together for a time the associated sand grains lying immediately below the upper surface while the lower portions of the cliff, relatively free from roots, and also subjected to a far more powerful sand blast owing to being nearer to the sand blasted surface, are more readily cut away by the sand blast. This undercutting of the faces of sand cliffs by the sand blast is well seen in Photo. 45. The undercutting is often assisted by the presence of rabbit burrows running beneath the surface.

As a result of the undercutting of the faces of the cliffs the upper portions of the cliffs which hold together for a time often come to overhang the basal portions very markedly. (See Photo. 45.)

Eventually however, when the undercutting has proceeded sufficiently, the overhanging upper portions break away and fall to the bases of the cliffs where they are ultimately disintegrated by the sand blast. Such debris of *Calluna* plants, etc., fallen to the bases of the sand cliffs and being gradually disintegrated by the sand blast, can be seen under the overhanging edges of the cliffs in Photo. 45.

Partly undercut *Calluna* plants on the upper overhanging edges of disintegrating undercut cliffs often hang suspended by their roots after some of the upper overhanging portions of the cliff have broken away. Sometimes these suspended *Calluna* plants live for a time in this position, but their suspending roots are eventually released by the continued disintegration of

F   6 the faces of the cliffs and the suspended plants eventually fall upside down on the sand blasted surface and the debris is ultimately disintegrated. In some instances a number of dead *Calluna* plants upside down and partly buried in sand with their roots projecting in the air can be seen on the sand blasted lower surface at a distance of several feet from the present face of a disintegrating sand cliff. In these instances no doubt these dead *Calluna* plants hung upside down suspended by their roots on the face of the cliff until these suspending roots were released by the continued disintegration of the cliff and the dead *Calluna* plants were deposited on the sand blasted surface in their existing positions before the face of the cliff retreated several feet to its present position.

Sometimes large rabbit-eaten *Calluna* hummocks on the edges of disintegrating sand cliffs bend over the edges of the cliffs as the latter become disintegrated, the formerly closely aggregated branches of the hummocks separating and showing fissures on the surface as the hummock bends over the edge of the retreating face of the cliff, as can be well seen in Photo. 46. No doubt the remains of the large rabbit grazed and rounded *Calluna* hummock seen in this photo will eventually fall upside down on to the lower surface and the debris will ultimately be disintegrated by the wind and sand blast. As mentioned above fragments of the roots and stems of disintegrated *Calluna* plants can be well seen on the lower surface of the cliff shown in Photo. 45.

In some cases *Calluna* plants on overhanging edges of the undercut cliffs die a long while before these undercut portions break away and fall to the bases of the cliffs, although all the other *Calluna* plants on the upper portion of the cliff are alive. Probably the porous soil and relatively dry climate of Breckland are in themselves not particularly favourable to *Calluna*, a plant of oceanic climates, and it appears very probable that the death of *Calluna* plants on the overhanging edges may very likely largely be due to exposure of many of the roots and reduction of the already small available water supply by the undercutting.

In some cases the main roots of *Calluna* plants on greatly undercut upper portions of cliffs reach down from these undercut upper portions and remain fixed in the surface strata below although the sand which formerly surrounded a great part of their length has been torn away by the sand blast. In these cases the *Calluna* plants on the overhanging edges often remain alive although they may be greatly undercut by the sand blast, and no doubt this is owing to the fact that, although they are so greatly undercut, they can continue to obtain a sufficient water supply through the roots, which, although partially exposed, remain fixed in the substratum.

It is interesting to note that *Calluna* stems on the immediate edges of greatly overhanging cliffs and now relatively inaccessible to rabbits are frequently comparatively tall and flower vigorously, while the stems of all the other *Calluna* plants in the vicinity, which are comparatively easily accessible to the rabbits, are nibbled closely down by them and are entirely prevented from flowering.

It has already been mentioned that the sand cliffs on the furthest edges of sand blasted areas tend to increase in height above the adjoining sand blasted surfaces both by removal of the original upper strata of these surfaces by the sand blast and by the deposition of some of the removed sand on the upper surfaces beyond. Such wind blown sand comparatively recently deposited upon the original surfaces beyond the edge of the cliff can be well seen on the exposed faces of the sand cliffs shown in Photos 45 and 46. In most cases these old surfaces were long occupied by vegetation (*Calluna*, etc.) and the strata just beneath them exposed on the faces of cliffs are usually much darker in colour than the strata of the comparatively recently deposited sand above them owing to being stained dark by the humus products of the original vegetation which existed for a comparatively long time upon them. (See Photo. 46.)

In consequence of this longer occupation of the original surface by vegetation the sand grains of the exposed strata lying just beneath the original surface are partly cemented together by the resulting humus products and thus these strata are considerably tougher than the strata of the comparatively recently deposited sand above them. As a result of this greater toughness of the strata lying just beneath the old surface, these strata when exposed on the faces of disintegrating cliffs often resist the disintegrating effect of the sand blast better than the more recently deposited upper strata of sand although they are nearer to the adjoining sand blasted surface. In consequence of this the lower strata (often themselves with undercut edges) frequently come to project further on the faces of sand cliffs than the more disintegrated upper strata of comparatively loose recently deposited sand, as can be well seen in Photo. 46. Thus the cliff may come to consist of two stages or tiers, each separately undercut.

Sometimes a particular upper surface of the comparatively recently deposited sand remains stable and uncovered by additional wind blown sand for a time, and when this happens—if the upper surface remains uncovered by additional sand for a sufficient period—it frequently becomes colonised by *Calluna*. Later on this *Calluna* which had colonised the relatively stable upper surface often becomes covered by subsequent wind blown sand, and ultimately the remains of it appear on the disintegrating face of the cliff as a thin stratum of decaying *Calluna* stems, etc., embedded in the sand at some distance below the subsequent upper surface, as can be well seen in Photo. 45.

Much of the sand which has accumulated at the edge of the sand blasted area is often carried up through "blow-outs" in the faces of the cliffs and deposited on the upper area. Blow-outs in the face of a cliff can be seen on the extreme left-hand side of Photo. 45. The face of the cliff seen in this photograph is gradually retreating across the downward sloping side of a slight valley and, largely in consequence of this downward slope of the surface, the cliff itself is gradually becoming obliterated.

In some instances the sand blast has already cut away nearly all of the former cliff in certain positions, except where portions remain as hillocks on the leeward side of masses of vegetation (principally *Calluna* hummocks) which have protected them from the sand blast. This effect of *Calluna* hummocks, etc., on the faces of disintegrating sand cliffs in protecting the portions of the cliffs lying just behind them from the disintegrating effects of the sand blasts reminds one of the action of *Suaeda fruticosa* bushes on mobile shingle banks in protecting the areas of shingle lying just behind them from the deposition of shingle in the rolling over of the surface of the bank during on-shore gales[1]. It is interesting to note that behind *Suaeda* bushes gullys are left whilst behind *Calluna* hummocks exposed to sand blasts hillocks are left. The sign of the effect of the protective action on the contours is reversed in the two cases. Much sand is usually carried along the blow-outs which exist between the protected remains, and becomes deposited on the vegetation beyond. Some areas of the sand blasted and deposited sand usually become colonised by *Polytrichum piliferum* except just beyond the blow-outs themselves where the surface is far too mobile for any sort of colonisation to occur.

Various phenomena strongly confirm the view that the faces of the relatively tall sand cliffs are homologous with the much more dwarf but frequently undercut faces of the much smaller sand cupolas and hummocks previously described[2]: the undercut faces of small vegetated cupolas and the faces of much taller vegetated sand cliffs are produced by the sand blast in the same manner, the one kind of projecting cliff face being a much larger variety of the other. The tendency towards the production of blow-outs through which the sand can escape in the case of long tall cliffs and the frequently resulting production of isolated portions of cliffs is probably the same tendency which results in the production of the small isolated cupolas around which the wind borne sand can escape, for the edges of a small cupola in the directions of varying winds may be regarded as the edges of an almost infinitely wide blow-out in the face of a dwarf cliff.

The sand carried up the face of the cliff is deposited amongst the vegetation on the higher ground beyond and any *Calluna* hummocks on this upper surface are often almost, and sometimes completely, filled with this wind blown sand. (Notice the sand in the *Calluna* hummocks on the upper portions of the cliff in Photo. 45.)

In the case of rabbit grazed *Calluna* hummocks this wind blown sand is first deposited on the windward side of the hummocks (see Photo. 45). This

[1] **Oliver, F. W.** and **Salisbury, E. J.**, "Vegetation and Mobile Ground as illustrated by *Suaeda fruticosa* on Shingle." *Journal of Ecology*, **1**, 1913, pp. 261-264.

[2] See Chapter VI; notably Photos 38, 40 and 41, Plates XIX, XX and XXI.

is interesting, for in the case of developing sand dunes the sand is chiefly deposited on the leeward side where the wind velocity is lowest. The difference probably arises because the *Calluna* plant has much denser vegetation than that of a developing sand dune, so that this denser vegetation can sufficiently reduce the wind velocity even on the windward side for the sand to be deposited there and remain at rest. The rabbit grazed and rounded *Calluna* hummock in Photo. 46 which is bending over the edge of the cliff as the latter gradually becomes disintegrated contains much wind blown sand.

Sometimes owing to the distribution of a sand blast tending to increase when once it has started great masses of sand are eventually cut away and extensive areas are bared. The general lack of vegetation on the surfaces in these cases is due to the surface sand blast and the mobility of the surface and is not due to excessive dryness, for these bare surfaces of sand are at a lower level and nearer to the underlying chalk than were the formerly stable upper surfaces which have been cut away and which were occupied by vegetation until the denuding sand blast cut them away. In the case of these large sand blasted bare areas the sand blasts sometimes become dormant for a time, and when this happens these formerly sand blasted bare areas gradually become recolonised. *Polytrichum piliferum* is the pioneer coloniser in these cases and is normally followed by *Cetraria aculeata, Cladonia coccifera* and *Ceratadon purpureus*.

It was thought that it would be interesting to have further information about the colonisation of bare sand and in order to obtain this the sand of a metre quadrat of grass heath was dug out to a depth of half-a-metre and removed and the space was filled with sterile sand carted from below a bare area. This quadrat for studying the colonisation of bare sand is inside the large rabbit-proof quadrat[1] and is protected by boards at its sides to prevent the loose sand from being blown away (middle quadrat in Photo. 31).

The bare sand in this experimental quadrat became colonised very rapidly by *Rumex acetosella* seedlings and a few months after it had been constructed there were very many of these on the quadrat. There were also some *Senecio vulgaris* seedlings and some wind blown *Cladonia* spp. on the sheltered edges near the boards. Several *Taraxacum erythrospermum* seeds were also found on this quadrat in the autumn after it had been constructed. In the spring there were many *Teesdalia nudicaulis* plants and the *Taraxacum erythro-spermum* seeds germinated and produced seedlings. Many *Galium saxatile* seedlings also appeared and being here protected from rabbits later on flowered profusely. One solitary *Cytisus scoparius* seedling appeared. Possibly the seed of the last named may have been carried and deposited by a bird.

It will thus be seen that the bare sand of this experimental quadrat is becoming colonised very rapidly. Most of the plants which are colonising this experimental quadrat are much taller than the dwarf mosses and lichens

---

[1] See Chapter IV, Photo 31, Plate XVI.

which are colonising the larger exposed formerly sand blasted bare areas already referred to. Probably the greater height of the plants on the experimental colonisation quadrat is chiefly owing to their being protected from attack by rabbits, but no doubt the relative absence of sand blasts may also have something to do with the matter. The bare sand of this experimental quadrat which is protected from rabbits is also becoming colonised much more rapidly than the exposed bare sandy areas. No doubt this is chiefly owing both to the absence of direct attack by the rabbits on the colonising plants and to the proximity of established plants protected from rabbit-attack inside the rabbit-proof enclosure which produce many more inflorescences and flowers than if they were exposed to rabbit-attack and can thus supply seed to the colonising area more quickly.

It has already been stated (see pp. 80 and 81) that when the sand blast has cut a cliff in the sand at the furthest edge of a sand blasted area, sand is usually lifted by the sand blast up and over the edge of this cliff and deposited upon the higher surface beyond. Sometimes fairly thick layers of sand are deposited in this way upon the upper surfaces, but sometimes the lifted sand is deposited as a thin mantle of sand over the vegetation on the upper surfaces. Frequently the vegetation of considerable areas is covered by a deposited mantle of sand in this way—sometimes for a distance of several hundred yards behind the edge of the cliff. These areas appear almost bare of vegetation for a time owing to the superficial deposit of wind blown sand.

In all of these cases however it was noticed that the areas on which the vegetation had been completely covered quickly became again occupied by vegetation—so quickly indeed that it seemed impossible to account for the reoccupation of these areas on any hypothesis of simple recolonisation.

In order to see more exactly what happened and in order to have definite information about the date of the deposition of the sand, a metre quadrat of grass-heath inside the large rabbit-proof enclosure was boarded off to prevent the sand from being blown away and sterile sand was then carefully sifted in and on the vegetation of this boarded metre quadrat to a depth of 5 cms. above the previous surface. (Left-hand quadrat in Photo. 31, Plate XVI.) The average height of the vegetation on this quadrat was about 2 cms. and its maximum height at that time was about 3 cms., so that all the grass-heath vegetation on this quadrat was completely covered by the 5 cm. deposit of sand and was mostly covered to a depth of 3 cms. In spite of this, the sterile sand covering the previous vegetation quickly became occupied by vegetation and at the end of two months the vegetation on the quadrat was quite as luxuriant as the vegetation on the untouched quadrat and many inflorescences had appeared. On examination it was found that the rapid recolonisation of the new sandy surface was owing to the vegetation which had been submerged by the deposit of sand rapidly sending up fresh shoots to the new surface and thus covering it again.

Many plants in Breckland have been found to possess this power of sending up shoots through a superficial deposit of sand and thus reaching the fresh surface. *Agrostis vulgaris, Festuca ovina, Festuca rubra, Galium verum, Rumex acetosella, Thymus serpyllum* and *Lotus corniculatus*, along with many others, were found to behave in this way.

The details of the process vary somewhat in the case of different plants. In the case of *Agrostis vulgaris* the main stem usually elongates, quickly grows up through the deposited sand and produces a fresh set of leaves at the new surface; but sometimes lateral stems also grow up, and adventitious roots are soon produced by the new stems. In the case of *Festuca ovina* several lateral stems usually rise through the sand deposit and produce fresh leaves at the new surface. These rising lateral stems often produce small etiolated leaves inside the deposited stratum and later on they also produce adventitious roots inside this stratum. *Lotus corniculatus* usually sends up a number of thin etiolated lateral shoots from the buried crown, and when these rising shoots reach the top they each produce fresh groups of leaves at the new surface and eventually adventitious roots are produced inside the sand. In the case of *Lotus corniculatus* these rising lateral shoots also produce etiolated leaves and etiolated secondary lateral shoots inside the sand stratum before reaching the surface.

It thus appears that many plants—by means of sending up fresh shoots to reach the new surface—possess the power of contending with such an apparently unfavourable environmental influence as being covered over with sand. *Ammophila arenaria* and other sand grasses have long been known to have very extensive powers of this kind, but it is becoming apparent that very many other plants have similar powers. *Suaeda fruticosa* also can contend with rising shingle and can rise through it by the production and growth of lateral shoots[1]. This method of contending with rising shingle is very similar to the general method by which all these other plants can contend with rising sand, and the general process is one of considerable interest.

This phenomenon may be rather important in connection with Richthofen's theory of the mode of origin of the loess[2], for this theory requires the subaerial deposition and stabilisation of ultimately considerable thicknesses of material. Since so many plants can rise up through subaerially deposited material, vegetation may well have greatly assisted in the subaerial deposition and stabilisation of ultimately great thicknesses. Probably the structures which Richthofen observed in the loess and termed "roots" were really continuous stems which had continually risen up through the successive strata of subaerially deposited material as it was deposited. Prof. Marr informed the writer that it did not seem that the long continuous structures

[1] **Oliver, F. W.** and **Salisbury, E. J.**, "Vegetation and Mobile Ground as Illustrated by *Suaeda fruticosa* on Shingle." *Journal of Ecology*, **1**, 1913, p. 259.

[2] **Richthofen, F.**, "On the Mode of Origin of the Loess." *Geological Magazine*, 1882.

which pass through considerable thicknesses in the loess could possibly have been formed by "roots" and this was one of the difficulties in Richthofen's theory of the mode of origin of the loess. If however these structures instead of being "roots" which had penetrated down were formed by continuous stems which had continually risen *up* through successive strata of subaerially deposited material, this difficulty in Richthofen's theory of the mode of origin of the loess would be explained.

In this connection much of the sand of Breckland—in any case for considerable distances behind certain sand cliffs—is a sort of subaerially deposited loess. Digging on many of these areas reveals stratified sections of the successively deposited sand layers frequently each about 1 cm. thick and long remains of the shoots with adventitious roots of various plants which have continually risen up through the successively deposited sand layers.

The vegetation of the quadrat upon which sand was deposited became quite as luxuriant and even closer than that of the untouched turf. The deposited sand probably had some bad effects on the vegetation but it seems to have had some good effects which have quite compensated for any bad effects. In this respect, these grass heath plants resemble many other plants such as *Suaeda fruticosa* and *Ammophila arenaria* at Blakeney, for these plants are often stimulated when shingle or sand is deposited around them. The effect is in some respects contrary to what might have been expected. It is not known definitely how this unexpected result is brought about, but perhaps it may be partly due to the older parts which have become buried deeper decaying and becoming available as manure for the new upper tissues which have been produced. It has apparently been found generally economical in nature for old tissues to die and to be replaced by new and vigorous tissue and it may perhaps be possible that the increased luxuriance which is often apparent in these cases may be partly due to a modification or speeding up of the life-death-life cycle.

There are doubtless many other instances in which plants contend with unfavourable environmental influences by producing new vertical shoots to rise to the new surface. The writer observed some good instances of this at Blakeney, where individual plants of *Glyceria maritima* were surrounded by a dense growth of *Obione portulacoides* which was gradually increasing in height. When this happened the *Glyceria* plants sometimes sent up new vertical shoots to the upper surface of the *Obione* layer, where they produced fresh sets of leaves which remained connected with the smothered leaves and the substratum by means of the new vertical shoots. These instances resemble in many ways the process by which other plants contend with rising sand, but in the case of *Glyceria* contending with rising *Obione* well-developed adventitious roots are not produced by the new *Glyceria* shoot as they are when a plant has risen through deposited sand.

CHAPTER VIII.

VIEWS RELATING TO THE PROBABLE FORMER DISTRIBUTION OF *CALLUNA* HEATH IN ENGLAND.

Calluna heath alternates with grass-heath in Breckland, and this was previously supposed to be probably due to varying proportions of lime in the associated soils—some of the grass-heaths approach chalk pasture in the character of their vegetation and the soils frequently contain appreciable quantities of lime.

As described in Chapter II, it was discovered however that many of these grass-heath areas arise by the degeneration of typical *Calluna* heath owing to biotic attack on the *Calluna*. This process of degeneration of *Calluna* heath to grass-heath occurs in many localities in Breckland and is a widespread phenomenon in this district. The process has probably been going on for some time and many areas now grass-heath were doubtless once typical *Calluna* heaths before they were rabbit attacked or grazed so heavily as they are at present.

Many things indicate that Higham Heath, which lies on the south-west border of Breckland, and which is now a pure grass heath, was very probably once a typical *Calluna* heath. Some plantations of gorse bushes have been made on Higham Heath but these are degenerating rapidly at their edges owing to the rabbit attack and to grazing by sheep. Some planted gorse bushes occur on Cavenham Heath and they do not degenerate so rapidly as the *Calluna*. Thus if gorse degenerates rapidly on Higham Heath, *Calluna* would have degenerated more rapidly, under comparable biotic attack. The absence of *Calluna* now is no evidence that it was never there.

Rumex acetosella, indicative of acid surface conditions, occurs on Higham Heath. *Calluna* is not intolerant of a high lime content below the surface layers of soil, provided other conditions are favourable, but heavy biotic attack by sheep or rabbits or both has been shown to be extremely antagonistic to its persistence.

Remarkable mixed communities termed "chalk-heaths" and "limestone-heaths" sometimes occur on soils overlying calcareous materials. These are characterised by a mixture of typical heath and typical limestone plants. The surface layers of undisturbed soils overlying calcareous rocks are typically extremely poor in lime although the lower layers of the soil are usually rich in lime. Various writers have recorded that *Calluna* can colonise the superficial layers poor in lime over calcareous rocks if the other conditions are favourable and it appears that these other conditions would include the absence of heavy biotic attack. If *Calluna* could once colonise the superficial layers, dead *Calluna* leaves would eventually fall on the surface and the typical subordinate vegetation of mosses and lichens would gradually decay so that eventually a considerable thickness of peat or peat-like substance would form over the superficial layers of soil overlying the calcareous rock, and typical *Calluna* heaths on superficial peat might eventually be produced, as is actually the case on many limestone and chalk plateaus.

Possibly extensive areas in England on certain soils may have been formerly occupied by *Calluna* heath after the degeneration of the primitive woodland from preliminary pasturing where the latter was not sufficiently severe to cause the *Calluna* heath type of vegetation to degenerate. "Chalk-heath" and "limestone-heath" communities may perhaps often represent attempts on the part of *Calluna* and its associates to colonise the most favourable portions of their former distribution area where surface layers are poor in lime when the intensity of the local biotic attack becomes temporarily sufficiently reduced.

It would be well to test this theory of the origin of "chalk-heath" and "limestone-heath" communities through locally reduced biotic attack by means of erecting grazing- and rabbit-proof enclosures in various places in such communities and noting the resulting effects. It may very likely be found that the protected areas will develop typical *Calluna* heath after the grazing pressure has been removed.

There are areas of grass-heath between Barton Mills and Newmarket the soils and grass vegetation of which resemble those of Cavenham Heath: fragments of species of *Cladonia* occur. It is very possible that these heaths were formerly *Calluna* heaths before the biotic attack of grazing animals was as heavy as it is at present.

There is in fact a series of areas called "heaths" running across this part of the country on the chalk of the "East Anglian Heights," viz. the Breckland Heaths (including Higham Heath), Newmarket Heath, and Royston Heath. These "heaths" become increasingly like chalk *pasture* as one passes from the Breckland Heaths to Royston Heath. "The tracts of country occupied by this (grass-heath) association are often called heaths, although the true heath plants (*Calluna*, *Erica*, etc.) may be entirely lacking, but all transitions are found between the grass-heaths and the *Calluna* heaths...." "This term ('heath') is even used in East Anglia for tracts of chalk pasture, e.g. Royston Heath, Newmarket Heath[1]."

When the primitive woodland degenerated (probably chiefly owing to early pasturing) large open spaces bare of trees resulted, and when these were uncultivated they were termed "heaths." What are now known as "heath plants," such as *Calluna*, *Erica* and their associates, were probably highly characteristic of these "large open areas bare of trees" (resulting from the degeneration of woodland) which were termed heaths.

[1] *Types of British Vegetation*, p. 95.

It is quite possible that Newmarket "Heath" and Royston "Heath" may once have been occupied by the typical heath vegetation on a layer of peat after the degeneration of the primitive woodland and before the biotic attack was as heavy as it is at present[1]. It is very possible also that many areas of the chalk where the superficial layers are poor in lime may once have been occupied by typical *Calluna* heath before the biotic attack was as heavy as it is now.

In this connection it is interesting to note that Mr A. G. Tansley and Mr R. S. Adamson have found *Calluna vulgaris* colonising, flowering and reproducing by seed inside one of their rabbit-proof enclosures on fairly typical chalk down at Ditcham Park on the Hampshire-Sussex border[2].

Mr Tansley has also informed the writer of some interesting phenomena which he observed in 1916 on the northern face of Moel Siabod in North Wales.

A stone wall runs along the mountain side at a height of 1700 feet above sea level. Below the wall there is a great deal of vigorous *Calluna*—indeed it is generally dominant—while in general there is no *Calluna* above the wall although the soil and slope conditions are just the same.

Many sheep are pastured on the area above the wall, which is a common grazing ground, while only a few sheep are occasionally pastured below the wall.

The stone wall has become ruined so that sheep can cross it in some places and there it has been replaced by a wire fence which runs in some cases a few yards above and in some cases a few yards below the ruins of the wall, with the result that the sheep are now stopped by the wire fence where the wall is ruined and by the wall where it is intact, while formerly they were stopped by the wall all along.

The effect on the *Calluna* is most striking. Where the wire fence runs *below* the ruined wall the *Calluna* between it and the ruined wall is in a bad way and is nearly dead, while the *Calluna* on the lower side of the wire fence is perfectly healthy (see Fig. 1). Where the fence runs *above* the wall so that the area between the fence and the wall which was previously exposed to sheep is now protected from them, scattered *Calluna* plants a few inches high and about three or four years old are colonising the ground and growing amongst the grass (see Diagram). This instance is an exceedingly pretty confirmation of the thesis that the existence of *Calluna* heath or grass-heath often depends simply on the relative severity of biotic attack, and fits in well with the theory that *Calluna* heath may often be and have been a par-

---

[1] In this connection it is interesting to note that one of the keepers on Newmarket Heath has subsequently informed the writer that there is a considerable amount of heather growing on "Long Hills" on Newmarket Heath, and that there was far more heather there 20 years ago than there is now. It is cut down badly by the mowing machine each year—unfortunately from its point of view before the seed is ripe. This keeper expressed the opinion that if it were not for this regular mowing there would be a great amount of heather on many portions of Newmarket Heath.

[2] See **Tansley**, "Early Stages of the Re-development of Woody Vegetation on Chalk Grass-land." *Journal of Ecology*, **10**, 1922, p. 173.

ticular developmental phase in the degeneration transition of woodland to grassland under the stress of increasing biotic attack. (See Diagram of the Biotic Zonation of Breckland, Chapter III, Fig. 6, p. 37.)

If the biotic attack was once low in intensity vegetation could decay on the surface and eventually produce surface layers of humus poor in lime which could become colonised by *Calluna* (cf. the gradual change from fen to moor formation consequent upon the production of surface layers of decaying humus poor in mineral salts).

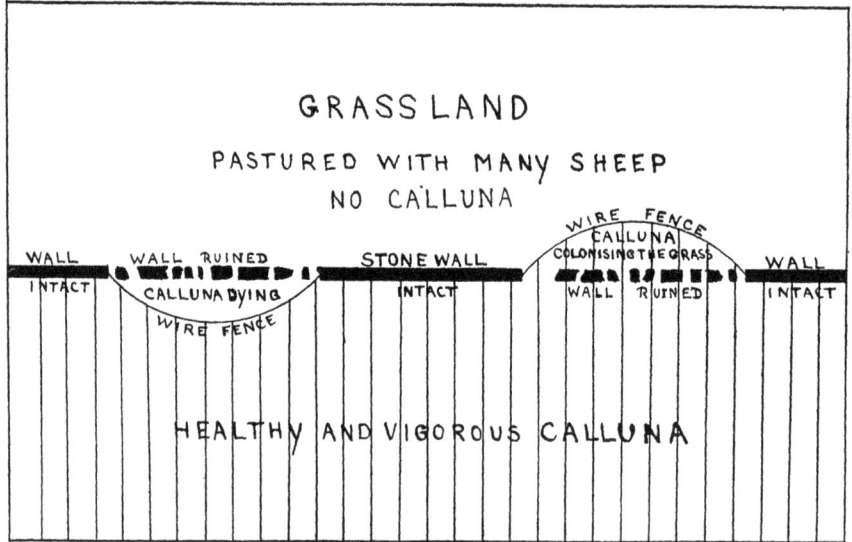

FIG. 10. Diagram illustrating the effect of sheep in limiting the distribution of *Calluna*. Moel Siabod in North Wales.

The writer is inclined to believe that as the primitive woodland degenerated, largely owing to early pasturing, typical heath plants (*Calluna, Erica,* etc.) frequently colonised the remaining surface layers of decaying humus when the pasturing intensity was low and that thus woodland was often followed by broad zones of "heath." As the better heaths became more heavily pastured than the poorer ones and as the better soils became cultivated, the distribution of the *Calluna* and its associates became more and more restricted owing to the heavy pasturing and the cultivation, until the present distribution of *Calluna* in Britain on the poorer uncultivated soils not heavily pastured is probably merely a remnant of its former distribution. Exceptions to this general process would occur where the former process of extension of the distribution of the heather at the expense of the primitive woodland is still going on at the present day on the poorer only slightly pastured soils—such as where the oak-birch woodlands are degenerating to heathlands occupied by *Calluna*. In comparatively recent times in many parts of England what is now termed the "common" used to be termed the "heath." The use of "common" lands for *grazing* purposes is interesting in this connection. The writer has grown *Calluna* luxuriantly on fairly rich alluvial soil.

It is unsafe to infer (as is often done) from the fact that heather in England is typically associated with very poor soil that therefore heather prefers or flourishes best or would naturally survive best on very poor soils. *Calluna* can probably grow better on soils considerably richer than those with which it is commonly associated at the present time, but it has been exterminated on the better soils owing to heavier grazing and clearing in order to cultivate the better soils.

The survival of *Calluna* only on the patches of better soil in the same pasturing enclosure (i.e. with the same pasturing intensity), while on the poorer surrounding soil it has been exterminated by the given intensity of the biotic attack and has degenerated to dwarf grass, has probably something to do with the explanation of the case of the apparently anomalous distribution of *Calluna* which has been described by Rayner and Jones[1]. If this be so the survival of the *Calluna* only on the patches of better soil is probably a particular instance of the important general phenomenon, of great biological interest, viz. that organisms withstand specific detrimental influences only when the other conditions are favourable and die when they are not (cf. the restriction of tree growth to the damper valleys by biotic attack on the seedlings, Chapter III, p. 33). Thus the influence of biotic attack on limiting the distribution of *Calluna* must be considered in addition to the lime distribution.

It may be that *Calluna* heath is largely to be regarded as a transition stage in the degeneration of woodland to grassland on certain soils where the biotic attack is sufficiently heavy to bring about the degeneration of the woodland and prevent it from regenerating, but not sufficiently heavy to exterminate the *Calluna* or to bring about the degeneration of the *Calluna* heath to grass-heath.

Probably after the degeneration of the primitive woodland and before the days of heavy pasturing by large domesticated herds and before the days of extended cultivation the distribution of *Calluna* in England was far more extensive than it is at present. It is noteworthy that it still occurs in every one of the Watsonian vice-counties of Great Britain (*Lond. Cat.* Ed. 10, 1908).

---

[1] **Rayner, M. C.** and **Jones, W. N.**, "Preliminary Observations on the Ecology of *Calluna vulgaris* on the Wiltshire and Berkshire Downs." *New Phytologist*, **10**, 1911.

CHAPTER IX.

VIEWS RELATING TO THE PROBABLE FORMER DISTRIBUTION OF WOODLAND ON THE BRECKLAND HEATHS AND ON HEATHS AND TREELESS AREAS OF OTHER PARTS OF ENGLAND.

As no traces of natural woodland have been found on the upper dry sandy areas of Breckland and as the low rainfall and sandy soil make this district the climatically and edaphically driest in England, it had previously been thought doubtful if these areas ever bore natural woodland, and it was supposed that the dry sandy heaths of this district might quite possibly represent primitive heathland which had never been colonised by trees[1].

On this account it was thought interesting to try to find out whether these dry Breckland sandy tracts are really probably primitively treeless. Information relating to this would be likely to have a bearing upon the probable former distribution of woodland in other parts of England. Considerable attention has therefore been devoted to obtaining information on this matter.

These Breckland heaths are the nearest approach in England to the North German heaths which have been described by Graebner[2]. The North German heath-area differs from the south-east European steppe region in possessing a damper climate not so intensely cold in the winter months and cooler in the summer without long continued dry periods. "The great heath formation of north-west Europe is typically developed on relatively poor sandy and gravelly soils whose climate is wetter than that which gives rise to steppe. The steppe climate is too dry for tree growth apart from local edaphic conditions but the heath formation exists side by side with woods, and in many cases has arisen as the result of the degeneration of woodland[3]." However, as stated above, it seemed that the dry sandy tracts of Breckland might very likely represent primitive heathland which had never been colonised by trees.

The profusion of flint implements of a great variety of different periods and their relative absence from surrounding districts indicates that Breckland was especially favourable as a habitat for prehistoric man, and it seems very probable that this may have been owing to its being dry, treeless or only thinly wooded and easily cleared of trees in primitive times, whilst adjacent districts were covered by dense impassable forest with thick undergrowth frequented by and capable of sheltering predatory animals.

The ancient linear defensive earthworks which occur in Breckland and to the southwestward, viz. the Black Ditches, Devil's Dyke, Fleam Dyke, Bran

[1] *Types of British Vegetation*, p. 107.

[2] **Graebner, P.**, *Die Heide Norddeutschlands*, Leipzig, 1901.

[3] **Tansley, A. G.**, *Types of British Vegetation*, p. 98.

or Heydon Dyke, etc., run from the fenlands near the river valleys across the Chalk escarpment up to the Boulder Clay, which was once occupied by forest. The fosse or ditch of these dykes is typically on the western side.

In all probability these dykes were arranged across what was open treeless country at the time of their construction in order to prevent enemy tribes from invading the country of the eastern tribes from the west and driving away their domesticated herds of cattle across the open space.

Thus the areas crossed by these ancient defensive earthworks were probably open and treeless or at most very thinly wooded at the time of construction of the earthworks, and if the date of construction of these earthworks was known, it might be safely inferred that the areas were treeless or only thinly wooded at that date, and that if they were ever previously covered by a fairly thick tree community, most of the trees had disappeared or been destroyed before that time. Unfortunately, however, very little is known about the age of these dykes [1, 2], except that Fleam Dyke is partly at least post-Roman. The most recent discussion [2] concludes that some at least are probably pre-Roman, and that the early Iron Age is a likely date for the Black Ditches (the only dykes of this series in Breckland proper), and for the Devil's Dyke, which crosses the chalk.

Most of Breckland was undoubtedly treeless during comparatively recent times until about the middle of the eighteenth century, when *Pinus sylvestris* was introduced into the district and various pine plantations were made on the open heaths [3].

These pine plantations were often called by such names as "Folly Wood," so—if one may judge from place-names—it may be considered that the inhabitants imagined that the venture of making pine plantations in the otherwise treeless Breckland was a foolish act and would prove a failure.

At the present time these pine plantations are fairly flourishing and the trees, judged from their size, appear to be about 50 years old. As a matter of fact in many cases they are more than twice that age. Thus the trees only grow very slowly and in all probability this is chiefly due to the very poor water supply.

In addition to the pine plantations, various artificial plantations of *Quercus sessiliflora* and *Betula alba* sometimes occur on the upper sandy tracts. The trees in these deciduous plantations are not well grown, but nevertheless they can and do grow to a certain size. It must however be remembered that in the artificial plantations the young trees may have been protected from the competition of the heather for the limited water supply.

Occasionally bushes of *Crataegus monogyna* occur amongst the heather,

---

[1] **Hughos, T. McKenny,** "The Cambridgeshire Dykes." *Journal of the British Archaeological Association,* **19,** 1913.

[2] **Fox, C.,** *The Archaeology of the Cambridge Region.* Camb. Univ. Press, 1923.

[3] **Clarke, W. G.,** "Some Breckland Characteristics." *Trans. Norfolk and Norwich Naturalist's Society,* **8.**

but in these cases the branches are sometimes stag-headed and the trees are usually definitely limited in size. This applies for instance to an isolated specimen of *Crataegus monogyna* known as "Mile Bush" on Cavenham Heath. According to information received from an old shepherd, this bush is over 100 years old and has remained of a constant size for the past 60 years.

Sometimes seedling pine trees and young oaks occur amongst the thick heather and seem to be able to compete successfully with the heather, for they do grow, although slowly. On the upper drier areas however, when exposed to rabbit attack, they are always killed off by the rabbits before they can reach a sufficient size to become relatively immune. When rabbits are present in large numbers young seedling pines, oaks and birches can only colonise the valleys and lower portions of the valley sides[1], probably chiefly because only in these damper positions can they produce tissue sufficiently quickly to overcome the damage caused by the rabbit attack. It has already been stated that numerous young pine trees are colonising and growing amongst thick heather inside a large rabbit-proof enclosure on the dry upper portions of Cavenham Heath near the Icklingham Road, although no young pines survive on the upper portions of the heath outside the rabbit-proof enclosure.

The area marked 15 on the map of Cavenham Heath (p. 6) was once cultivated and protected by a rabbit-proof fence according to information received in Cavenham, and when the cultivation ceased it quickly regenerated *Calluna* heath. It was protected from rabbit attack for a time and the heather quickly became colonised by young pines. At that time some sheep also used to have access to the general area of the heath, and when the protecting fence round this area was eventually removed, the young seedling pine trees which had colonised it were quickly killed by the attacks of the sheep (and also probably by the rabbits)[2].

It is thus seen that the rabbit attack and the previous sheep attack limits tree growth to the valleys and prevents the upper dry sandy areas of the heaths from being colonised by young trees and from being ultimately converted into woodlands.

Various observations indicate that a kind of plant which can grow up above another kind of plant has thereby a very great advantage in competition with the more dwarf vegetation and will under ordinary conditions dominate the latter, but that on the other hand the taller plants tend to suffer more from, and may be exterminated by, biotic attack.

From all these facts it appears that although the conditions in Breckland are the climatically and edaphically driest in England, *Pinus, Quercus,* and *Betula* seedlings can colonise and grow in thick heather when protected from rabbit and sheep attack and also in plantations. The ultimate "natural"

[1] Chapter III, p. 33.

[2] It may be noted that the effect of rabbits on young trees is apt to be especially bad when the ground is covered with snow to such a depth that little but the stems of the young trees project above the snow mantle and are thus the chief kind of vegetation accessible to the rabbits.

highest type of vegetation on these sandy Breckland tracts under the present climatic conditions and in the absence of such animals as rabbits and sheep would thus probably be a pine or oak-birch woodland, but the presence of rabbits and the previous presence of sheep have kept down the taller growing trees which would otherwise cover the area, and have allowed the shorter growing heather to survive.

Rabbits have only been introduced into England in comparatively modern times and their natural enemies have only been systematically kept down very recently. In Palaeolithic times domesticated herds do not seem to have existed and other grazing animals were probably kept down to relatively small numbers owing to the various predatory animals which existed at these periods. Breckland was undoubtedly partly surrounded by forests which could also supply tree seeds and thus it appears that during any periods in Palaeolithic times in which the climatic conditions resembled, or were damper than, the existing ones, Breckland was probably occupied by woodland provided that grazing animals were not numerous and provided that Palaeolithic man did not destroy all the trees. If the grazing intensity was very low and tree felling scarcely existent, it might even have been occupied by woodland during any Palaeolithic periods in which the climatic conditions were somewhat drier than they are at present. These remarks would also apply under these conditions during various periods in the Neolithic epoch, but during the later periods anyhow, and very likely during the early periods as well, tree felling by man was extensive, and man was fairly well equipped for dealing with predatory animals. Herds of domesticated animals also existed during these later periods, whereas there is no evidence that domesticated animals protected from their natural enemies existed in Palaeolithic times, during which period grazing animals were probably comparatively scarce.

It thus seems that when the grazing intensity was low and tree felling not extensive, Breckland was probably wooded, but that when tree felling was extensive and herds of domesticated cattle protected from their natural enemies existed, the probably former existing primitive woodland degenerated owing to felling of the old trees to clear the ground, and to provide wood for various purposes, and owing to the attack of the domesticated herds upon the tree seedlings in preventing them from growing and rejuvenating the woodland. It appears that the probable effect of prehistoric man and his herds in destroying the primitive woodland and killing the struggling young trees was continued later on into historic times by flocks of sheep pasturing on the heaths, and still later by the attack of rabbits after these latter had been introduced into England. Thus after any ancient woodland on this area had degenerated long ago owing to human occupation of the district and attack of domesticated herds, woodland never got a chance to regenerate and so an almost pure heath association was left behind from early Neolithic times until modern times when plantations protected from grazing were made on the heaths. It thus seems that if the biotic attack was slight, as it probably was, Breckland was probably wooded in early times.

As a matter of fact at the time of the Norman Conquest there was woodland in various places on the N.E. corner of Breckland, and this woodland has since degenerated to heather. "There was woodland at various places in the eastern part of Breckland, Merton alone providing pannage for 240 hogs. During the next three or four centuries most of the local deeds contain reference to 'bruaria,' that is unproductive ground covered with heather and gorse[1]." These woodlands which existed in historic times were however on the extreme borders of Breckland[2].

The young pine seedlings and the pines in the plantation seem to be able to grow better than the oaks and birches and this indicates that if *Pinus* was present in the district at any particular period during ancient times when the grazing was slight and when the climatic conditions resembled the existing ones it would probably have been able to compete successfully with any oak-birch woodland which might have existed, and would have been able to establish pine woods on these areas. *Pinus sylvestris* seems to have been abundant throughout Britain during part of the Neolithic epoch[3], and if it had been present in Breckland during any particular period when the climatic conditions resembled the present ones any primitive woodland would probably chiefly be pinewood; but if *Pinus sylvestris* was absent from the district (as it seems to have been later on before it was reintroduced) any primitive woodland would probably consist chiefly of oaks and birches.

Various evidence seems to indicate that subsequently to the great Ice Ages, steppe periods may have intervened between periods of decreasing glaciation. If this be so, the gradual approach of steppe conditions might have helped in the degeneration of primitive woodland to an almost pure heath association. If any intervening steppe periods occurred during which the climatic conditions were much drier than those obtaining at present possibly these areas may have been too dry for tree growth even without grazing, and possibly any other vegetation may have been very sparse. These conditions would have facilitated the extensive blowing of sand which seems to have occurred in this district.

It has been considered that "natural causes of the degeneration of woodland and other associations are for the most part little understood[4]," but the explanation above given of the degeneration of probable primitive woodland

---

[1] **Clarke, W. G.**, "Some Breckland Characteristics." *Trans. Norfolk and Norwich Naturalist's Society*, **8**, p. 557.

[2] In passing it may be interesting to note that woodland on the edge of Breckland was the scene of the probably actual tragedy, which gave rise to the English version of the *Babes in the Wood* form of Fairy Tale. See **Kent, Chas.**, *The Land of the Babes in the Wood*. Jarrold and Sons. (The house of the "Wicked Uncle" is still standing.)

[3] **Reid, Clement**, *The Origin of the British Flora*. Dulau & Co.

[4] *Types of British Vegetation.*

in Breckland to a heath association owing to the attack of domesticated herds upon the seedlings in addition to the clearing by primitive man, and the prevention of the regeneration of the woodland in modern times owing to the attack of rabbits, accords fairly well with the explanation given by Krause and Borgreve for the degeneration of oak woodland to *Calluna* heath in North Germany, and it is not in accordance with the theory of leaching of sandy soils given by Graebner for the same area.

In any case these Breckland sands have not been leached sufficiently by rain for tree growth to be unable to compete with the heather, for tree growth does as a matter of fact occur—in plantations and in enclosures where the young trees are protected from rabbits. The heather itself is usually thicker inside these enclosures and yet the young trees grow much better amongst thick heather when they are protected from rabbit attack than amongst thin heather where they are exposed to rabbit attack. This shows that the pines can compete satisfactorily with the heather and that leaching of the porous soil has not rendered the area incapable of becoming colonised by pines or brought about the degeneration of possible primitive woodland, but that this may have been brought about by biotic attack.

Graebner's theory of leaching is also in some other cases not satisfactory to account for the degeneration of woodland. In fact one would not expect leaching of the soil to counterbalance readily the great advantages which the tall *Pinus* would have over the dwarf heather, but grazing or rabbit attack can counterbalance this—in accord with the generalisation that taller growing plants are usually the ones which ultimately suffer most from increasing intensity of biotic attack. The point here is the *relative* effects of biotic attack of increasing intensities. A relatively slight intensity of biotic attack puts the trees at a great disadvantage compared with the dwarf heather, and ultimately when the biotic attack becomes much greater, as in various places in Breckland at the present time, it eventually leads to degeneration of the heather itself, which becomes replaced by *Carex arenaria* and ultimately by still more dwarf grass-heath.

Graebner attempts to prove the poverty of the soils occupied by heather relatively to the soils occupied by forest by data obtained from chemical analysis of the soil, especially from those made by Ramann. These results support the view that heaths occupy the poorer soils, but this fact is no proof whatever that the heath has come to occupy the soil owing to the soil being rendered poor by leaching. The occupation of the poor soils by the heather may be, and probably is, due to quite a different cause, viz. destruction by biotic attack of its taller competitors which were formerly smothering it, a destruction which would not occur so easily on the better soils with the same intensity of attack.

Graebner's theory of woodland degenerating to heath owing to leaching of the sandy soil may indeed be largely the opposite of the truth, i.e. an effect instead of a cause, i.e. the poverty and leaching of the sandy soil may be largely a result of the degeneration of forest to heather by the attack of pastured herds and rabbits, thus exposing the soil to the direct leaching effect of rain, instead of the degeneration of the forest being due to leaching of the sandy soil. When the forest is degenerating owing to these other causes *Calluna* is able to invade its degenerating edge, and later on when the trees have disappeared the water which would otherwise be absorbed by the tree roots may soak through. But in the case of Cavenham Heath and other Breckland Heaths this has apparently not leached the soil sufficiently to prevent the tall growing trees from having a great advantage over the relatively dwarf heather and regenerating the forest when once the biotic attack is again removed, as inside enclosures.

It is apparently very dangerous to make deductions from soil analysis alone, as the particular differences in the soils may often largely be due to the presence or absence of particular plants instead of the absence of particular plants being due to particular differences in the soils.

It has already been mentioned that the fact that *Calluna* heath degenerates sharply at its edges when the degeneration is caused by biotic attack due to animals requiring extra food is suggestive, since forests and various other associations also chiefly degenerate at their edges.

It is sometimes stated that when the whole of the ground vegetation is of the heather type in degenerating woodland, it is doubtful if the young trees can grow[1]—though it is not implied that this is necessarily due to the heather itself. But the growing of young trees amongst thick heather in rabbit-proof enclosures on Cavenham Heath and their inability to grow amongst thinner heather when exposed to rabbits indicates that for Cavenham Heath this inability of the young trees to grow is not due to the whole of the ground vegetation being of the heather type but to the rabbits or other grazing animals which have probably originally brought about the degeneration of possible primitive forest killing off the young trees very rapidly indeed.

It is probably incorrect to speak of the taller woodland "giving way" owing to invasion by heath plants, as is often done. The woodland may "give way," but this is not owing to invasion by the lower growing heath plants. The invasion of the woodland by the heath plants is due to the "giving way" of the woodland owing to biotic attack on the seedlings.

The surface layer of dry peat and other things may render the young trees more readily killed by the rabbit attack, but in the case of Cavenham Heath and other Breckland Heaths rabbit attack is undoubtedly the immediate and effective factor in the matter and if it were not for this the *Calluna* heaths would regenerate to woodland.

The degeneration of woodlands to heather wastes of little value has been a serious economic problem in Germany and it appears probable that the

[1] *Types of British Vegetation*, p. 100.

practically valueless heather wastes of Breckland might be converted into useful pinewoods and perhaps dry oakwoods if the numbers of the rabbits were reduced and grazing on the open heaths was avoided. The grazing is not very profitable since the land is so poor and animals do not like the heather. The most important things which they eat are probably the young seedling trees which would otherwise gradually colonise these areas and change them into useful woodlands, and the afforestation of the heather wastes of Breckland would provide a certain amount of useful employment.

The slight humus and iron pans which have already been described as occurring in these heathland soils are certainly not sufficiently formidable seriously to hamper the growth of trees, and if these areas were afforested, there would be no need to break up these slight pans.

BEARING OF THE ABOVE FACTS ON THE PROBABLE FORMER DISTRIBUTION OF WOODLAND IN OTHER PARTS OF ENGLAND.

Very little semi-natural woodland is associated with the heaths of the London basin and it seemed possible that the heath formation originally colonised this and many other areas of the poorer English sands such as the Pliocene Crag, and in the south possibly also the Eocene Bagshot sands, and that these sands, or at any rate parts of them, had never borne natural woodland[1].

The writer has, however, observed young seedling trees colonising the upper portions of the Eocene Bagshot sands and many other areas of English heaths and this phenomenon indicates that these sands can support tree growth and that they are not too dry or poor for this, and also that any original degeneration of primitive woodland on them has not been brought about by the soil becoming leached so much that it has become too poor for the trees to compete satisfactorily with the heather. It seems probable that any original degeneration was brought about by grazing and clearing and that biotic attack (under the already unfavourable conditions) has kept them comparatively free from the surrounding tree growth ever since, as in the case of the Breckland heaths.

In some cases such as the heaths of the London basin it seemed impossible to decide whether the heaths were primitive or derived from woodland—in other words to draw the limit between possibly primitive heaths and heaths which had undoubtedly been derived from woodland, for the two possible cases might be represented at the present time by identical plant associations[2].

Since however it appears that the dry and poor sandy heaths of Breckland, which are climatically and edaphically the driest in England, can support trees provided that the seedlings are protected from biotic attack, and were

---

[1] *Types of British Vegetation*, p. 99.
[2] *Ibid.* p. 10.

probably wooded in pre-neolithic times when grazing animals were probably relatively scarce before the existence of domesticated herds protected from their various natural enemies, the apparently difficult problem of deciding whether any particular English heath was primitive or derived from woodland may have been indirectly solved. It seems that if grazing animals were once relatively scarce all the other English heaths probably once supported woodland and that probably none of them are primitively treeless.

It has also been considered that much of the area of the chalk downs has possibly never been covered by woodland. "There is good reason to suppose, as we have already pointed out, that much of the chalk pasture is extremely old and much of its area has possibly never been occupied by a tree association—perhaps because of an inadequate supply of underground water. The chalk grassland, which forms a very excellent light crisp pasture, has from time immemorial supported considerable flocks of sheep.

"The smooth curves of the chalk downs are occasionally broken by ancient trackways, camps and other earthworks of many periods from the Neolithic onwards. It has been suggested that the original purpose of many of these works was to shelter and defend the flocks from the attacks of predatory animals such as wolves coming from the forests of the lower country. Be that as it may, it seems unlikely that primitive man was responsible for the disforestation of such great areas of the chalk upland as are marked by traces of his presence and the conclusion is therefore indicated that much of this grassland is primitive, or at least has existed since the conditions of climate resembled at all closely those at present obtaining. There may well have been originally more scrub than there is now[1]."

The fact that the young trees degenerate so rapidly and readily from biotic attack on the upper dry sandy areas of Breckland where the conditions are already very unfavourable for them, whereas they do not degenerate nearly so readily from this influence and often manage to survive down the valley sides where the conditions are more favourable, is very interesting in this connection, for on the above mentioned areas of the chalk downs the conditions are probably also relatively unfavourable for tree growth. Thus the young trees on these areas of the chalk downs might degenerate very rapidly and readily from biotic attack by grazing animals compared with the rapidity of biotic degeneration under more favourable conditions. It thus seems that the above-mentioned vast tracts of chalk pasture on the downs may quite well have degenerated from primitive woodland owing to grazing especially if the Neolithic epoch was one of very vast duration as other evidences tend to indicate was in all probability the case.

The fact that the grass vegetation on the chalk downs grows considerably more luxuriantly than that on the Breckland heaths when both are protected from rabbits, probably indicates that there is a greater supply of underground

---

[1] *Types of British Vegetation*, p. 173.

water on the chalk downs than on the porous sandy tracts of Breckland, and yet even the latter would become colonised by trees if it were not for biotic attack on the tree seedlings, or if the biotic attack on the seedlings was low in intensity[1].

The very rapid degeneration of seedling trees through biotic attack under the other unfavourable conditions in Breckland is also interesting in view of the fact that the conditions above the present altitudinal forest limit are also relatively unfavourable to growth, and it may well be that the extent of the depression of the altitudinal forest limit by the grazing of animals pastured above the tree zone may have been very considerably greater than might otherwise have been thought to be the case.

From observation in Breckland it appears that, if grazing animals were once relatively few before the existence of domesticated herds protected from their natural enemies, the primitive distribution of woodland in England under the existing climatic conditions was probably considerably more complete than has sometimes been supposed and that probably all the heaths of England[2] along with various other areas now treeless once bore woodland.

[1] Mr Tansley has pointed out to the writer the possibility that the human and grazing factors preventing tree growth may quite conceivably have become operative before the general invasion by trees of the poorer and drier soils, such as those of Breckland and parts of the chalk downs, was completed; and this is the more likely if dry continental conditions intervened between the final retreat of the ice and the prevalence of climatic conditions like those now obtaining. If this were so we cannot altogether exclude the hypothesis of primitive treelessness of some areas, although the writer thinks that the presence of trackways, implements and other signs of early human habitation of such areas may quite well mean that man was also responsible for the disforestation of these areas.

[2] Except perhaps those on coasts exposed to violent winds.

CHAPTER X.

CONCLUDING REMARKS UPON BRECKLAND AND UPON ECOLOGICAL RESEARCH IN GENERAL.

ONE thing which clearly emerges from the work recorded in the foregoing pages is the great importance which must be attached to the biotic factors of the environment. Ecological factors are sometimes grouped solely as edaphic and climatic, but the highly important biotic factors should probably always be included in the classification. Apparently the presence of rabbits alone is sufficient to change the potentially dominant plant on Cavenham Heath from *Pinus sylvestris* to *Pteris aquilina* through a large number of various stages. This is clearly a profound change. The passing of England from a forest period into a grassland period may of late have been accelerated by the influence of rabbits.

While producing great changes in the ultimately dominant types of vegetation, the differential effects of particular intensities of rabbit attack upon the various types of vegetation produce great movements in the respective distributions of the different vegetation-units even when regarded over comparatively short periods of time. In fact one of the chief characteristics of the vegetation of Cavenham Heath is the extreme mobility of the various types owing chiefly to the differential influence of rabbits upon them.

The vegetation of Blakeney Point is also considerably more mobile than might have been suspected[1]. Doubtless the alterations in the vegetation due to topographical changes are far more rapid at Blakeney Point than on Cavenham Heath, owing to the presence of mobile shingle banks and sand-dunes in the former case; but, on the other hand, the general changes in the vegetation are far more rapid on Cavenham Heath than on Blakeney Point owing to the greater operation of biotic factors on Cavenham Heath, which completely outweighs in this respect the comparative absence of topographical changes. Breckland is a very suitable district in which to realise the great effects of animals upon plant life—owing to its uncultivated nature, the variety of its vegetation and the increasing severity of the grazing influences. Although the chief results are probably capable of very general application, the writer is sceptical as to whether there are many other districts in England which show such an extensive and beautiful generalised degeneration and zonation round the rabbit burrows, passing through so many stages as dealt with in Chapter III.

Ecological research is concerned with living things, as they exist and

---

[1] **Oliver, F. W.** "Blakeney Point in 1914." *Report of the Local Committee of Management, National Trust.*

change and interact in the living world, and it is an extremely interesting and fascinating study. One of its merits is that it gives the mind some faint conception of the complexity of nature. The problems awaiting solution are innumerable and many of them are extremely difficult to attack successfully. "The chief obstacle to the rapid development of ecology on fundamental lines is the laborious and time consuming nature of the work[1]."

Owing to the difficulty and complexity of the problems involved, it is advisable to choose the definite field problems which are to be taken up with very great care as otherwise much time may be wasted in futile attacks delivered along wrong lines.

It is probably much better in most cases to attack the easier problems or positions first, in the hope that these may throw some light upon or command the more difficult problems and positions and, as it were, outflank them. This is probably a better method and will ultimately advance the whole of ecology more than the method of devoting all the energy simply to a frontal attack upon some individual problem. It is probably advisable to devote much attention to the strategy and tactics of the attacks before delivering them with energy. This method of working will probably give far more results than a series of attacks delivered without careful arrangement. It is probably advisable to follow the Baconian method in the earlier stages of a research and to observe and collect a large amount of data relating to various phenomena before spending a lot of mental energy in trying to invent hypotheses to explain the various problems. While one is collecting preliminary data many of the problems in one's mind resolve themselves and the data themselves almost automatically suggest to the mind likely hypotheses to explain the more difficult, and as yet unsolved, problems. Various experiments should then be carefully devised to try to test the various theories. Experimental methods used upon the facts of nature mark a great advance in checking and confirming the results of observation and inference and in giving fresh reliable results and sources of fresh data. Experiments tend to give definite results whereas soil analyses, for example, used alone in conjunction with possibly erroneous deduction are apt to give results which may very likely be the opposite of truth, for instance the "effect" instead of the "cause."

Instead of attempting to deal with ecological problems in the laboratory where very many factors are varied from their natural values, so that they are liable to alter or modify the result of the experiment unknown to the experimentalist, it is probably much better to take experimental methods to the plant in its home. Ecology means the study of organisms in their homes and the various theories relating to the actual ecological problems should be tested by experiments in the field itself, only altering one test factor at a time, and thus getting results definitely related to the effect of that factor upon the organisms, all the other factors remaining at their natural values.

[1] **Tansley, A. G.** *Types of British Vegetation,* p. xi.

Some of the problems which now badly need solution are those relating to competition between plants. Very little is known about this at the present time, and the solution of some problems relating to competition would be very valuable as being likely to throw important light upon many problems. At the present time the term "competition" is frequently used to cover any phenomenon which is not understood, and if more were known about competition, important light would probably be thrown upon many phenomena which are at present mysterious.

The observation of the great differential effects which the ravages of rabbits exert upon competing types of vegetation is possibly an advance in the right direction and may illustrate possible effects of the ravages of war upon human societies, but an enormous amount of work remains to be done, and should be done, upon phenomena associated with competition.

The effects of biotic influences by man and grazing animals upon the vegetation of England have probably been far greater and more extensive than is commonly or frequently realised—for instance, if it were not for these influences large areas of England including the whole of Breckland, the whole of the North and South Downs, and a considerable zone above the present tree limit (where there is sufficient foothold for tree growth) would probably be covered by trees. Thus, when anybody is working at the ecology of any ordinary non-wooded, non-coastal, and non-aquatic portion of England, it is probably advisable for him to realise very clearly that it was perhaps once covered by trees and perhaps would be at the present day if it were not for these biotic influences. This attitude would probably tend to make the student chiefly interested in the dynamic aspects of the vegetation and in its changes during various periods of time rather than in merely describing its static aspect at any given moment.

INDEX.

For EU product safety concerns, contact us at Calle de José Abascal, 56–1°, 28003 Madrid, Spain or eugpsr@cambridge.org.

www.ingramcontent.com/pod-product-compliance
Ingram Content Group UK Ltd.
Pitfield, Milton Keynes, MK11 3LW, UK
UKHW060312090126
466816UK00021B/447